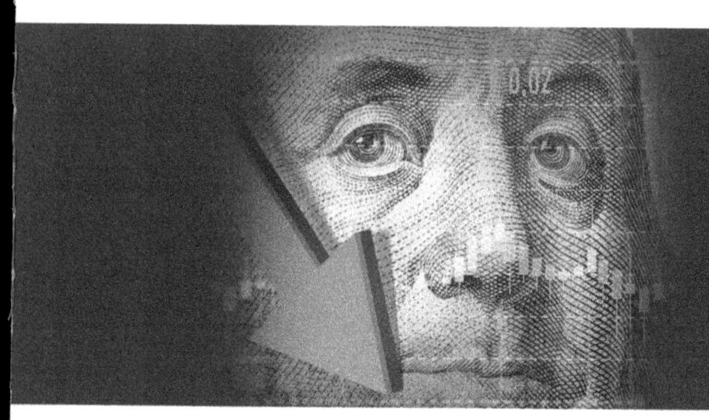

COLLAPSE

BY

WES MOORE

LEARN MORE AT

WESMOORENOW.COM

Published in the United States by Provectus Media. ISBN 978-1-63795-986-2.

Printed in the United States of America. FIRST EDITION. Cover and layout by Wes Moore. Edited by Sarah Crewey.

Wes Moore's other books:

Saving the American Church
Saving the Traditional Southern Baptist Church
Basic Bible Christianity
The Spiritual Top 50
The Maker

For more information, visit www.wesmoorenow.com.

TABLE OF CONTENTS

(Continued on next page)

TABLE OF CONTENTS (CONTINUED)

Introduction

Collapse. It's one of those words that is almost never used in a good sense. When a thing collapses, something valuable has been destroyed. Something good. Something important. When I hear the word, I see walls buckling, timbers splitting, roofs caving in. I see sad, broken people standing by in horror, tears overflowing their eyes. In short, I see death.

It's not a word you want used to describe anything you love. You never want it said that your marriage collapsed, your family collapsed, or your business collapsed. If you can make it through life without this term applying to you in any significant way, you have seen good days indeed.

But what about your nation? Think of the terror of the words, "America collapsed." It is too awful a statement to even consider, isn't it? But the way things are going, it seems we are destined to hear them anyway at some point.

Many of us grew up in a vastly different America than the one that exists now. There was a time when our nation, by almost any measure, was a land on the rise. We were a good people, a wise people, a people with two feet solidly planted on the ground.

But that time has passed.

We've lost our footing in America in a number of ways. We've lost it morally. We approve of and adopt increasingly godless ideas and practices in all facets of our society. We've lost it spiritually. We have rejected the Christian faith, the faith that has been the cornerstone of our nation, and have accepted any and all impostors in its place. And we're about to lose it economically. With greater and greater zeal, we abandon the free market, capitalistic ideas that made us what we are today and embrace the failed, poverty-creating doctrines of socialism.

This book is an attempt to address the latter of these three, the economic perils facing our nation. As socialism grows in popularity, its influence can be seen in more and more ways. Its effects are so subtle, however, that many Americans, especially believers, are totally unaware of their existence. Catchy phrases like "tax the rich," "profit is evil," "living wage," and "price controls," that seem so popular today, spring directly from socialist ideas, but we do not recognize them for what they are. And, when we do recognize them as socialist, we do not understand the destruction a full-scale implementation of this type of economic system would bring to our nation.

In the chapters that follow, I will tackle this problem from both an economic and biblical perspective. In so doing, I will reveal not only what sound economic principles tell us about socialism's chances for success, but also what the Word of God reveals on the issue. The Bible addresses almost every economic decision the nation faces, and it is precisely because we have abandoned its teachings that we are in the situation we find ourselves today.

But this book will do more than help you understand the basic principles of capitalism and socialism. It will also lead you to a more important realization, a realization that our nation, because of its financial and economic decisions, is headed for a collapse that could make the Great Depression seem like a boom time. As the book develops, the threads that are the individual chapters will come together to reveal the single issue that could bring down the greatest nation the world has ever seen—debt.

DON'T FEAR—YOU CAN UNDERSTAND THESE THINGS

Before I close these introductory remarks, let me address a fear that may be rising in some. If you are new to the subject of economics, you may fear you cannot understand it. I know exactly how that feels. When I first encountered the topic, it was confusing to me as well, but that was mostly because of how it was taught. In this book, I try to overcome the failings of economics textbooks and teachers by explaining things in a way that is easy for the average person to understand. I avoid the big terms that are the focus of most studies of economics, and I never use a single formula or technical graph.

What you will learn in this book is the basic stuff you've heard about all your life but never understood, presented in a way that will educate and empower you. My message to you, then, is simple: don't be intimidated. You *can* understand this material, and you will be *blessed beyond measure* if you do.

My hope is that the teachings of this book will awaken my fellow Christians to the foolish ideas we are implementing and, by doing so, become an instrument by which God may save this fantastic land I call home. May the Creator of all things, the great God of the Bible, be pleased to do just that.

In the brotherhood of Christ,

Hex Moore

CHAPTER ONE: WHY GOD CARES ABOUT THE ECONOMY

Does the title of this chapter surprise you? Does it seem strange that I would suggest that God cares about something as time-bound and secular as an economy? After all, God cares about spiritual things, like salvation, discipleship, and truth, right? He doesn't bother himself with human concerns like unemployment, wages, and interest rates, does he?

Yes, he does. He cares very much about these things, and his Word addresses economic issues in many places. But that begs the question, doesn't it—*why* does God care? Great question. But before we talk about that, let's define what an economy actually is.

WHAT IS AN "ECONOMY," ANYWAY?

The word "economy" has to do with getting the most out of what you have. It means being thrifty or efficient with something. Sometimes we even use it to mean a low-cost version of a thing, like an "economy car," for example. When applied to a nation, the word has a similar meaning.

> **WHAT IS AN ECONOMY?**
>
> A nation's **economy** is its way of satisfying the endless needs and wants of its people with the limited resources it has.

A nation's economy is its way of satisfying the endless needs and wants of its people with the limited resources it has. Needs are the things we must have to survive; wants are the things we'd like to have if we could afford them. (From now on, I'll use the word "needs" to cover both wants and needs, even though they are technically different.) Resources, on the other hand, are the things we have available to meet those needs, like money, farmland, microchips, and labor.

A key part of the definition of an economy is found in the word "endless." If you think about it, our needs really are endless. Consider the needs of every person in America. Without the barrier of what something costs or how much is available at a given time, is there really any limit to what we would have for ourselves?

Think about yourself for a moment. If you could make a list of everything you needed with no boundaries, would that list ever end? No. As soon as you thought of one thing, you would think of another and another and another. Now, multiply that by 300 million, and you can start to see how big the needs of a nation can be.

But, as we all know, there are barriers keeping us from having everything we want. Most of the time (in Western nations, at least) the biggest barrier is price. You only have so much money, so you must prioritize what you buy. There is also the limit of what is available. You might want something that no one makes anymore or that is out of supply at that moment. These realities limit what you can have.

A nation deals with the same kinds of limitations, hence the second important word in this definition, "limited." There is only so much money to go around (economists call this the "money supply"). There are also only so many factories, workers, tools, and farms. Consequently, there are only so many milkshakes, tablets, and swiss cake rolls to go around (and whatever else we might need). Given these limitations, the people of a nation must decide what they will make, how much they will make,

FIGURE 1.1 - THE GOAL OF ECONOMICS

Economics is about meeting unlimited needs (left) with limited resources (right).

how much it will cost, and who gets it. The decisions they make regarding these things determines what kind of economy, or economic system, they will have.

There really are only two types of economic systems, socialist and capitalist (or some mixture of the two). I'll have much more to say about both of them in chapter 2, but for now let me simply point out that not all economic systems do as good a job at supplying the needs of people as others. Some lead to abundance, opportunity, and freedom, while others lead to poverty, despair, and oppression. And this is where we start to see why God cares about these things. God cares about *the economy* because God cares about *people.*

But we are getting ahead of ourselves again. Before we discuss how economic systems affect the lives of people, let's review some key components of economies and some common terms used when discussing them.

PARTS OF AN ECONOMY

Economies are made up of markets. Markets are places, either real places (like a hardware store or farmers' market) or virtual places (like Amazon.com or eBay) where buyers and sellers come together to exchange products for money (see Figure 1.2). Buyers, often called "consumers," are the people who have a need for something; sellers, also called "producers," are businesses that provide things to meet needs. The money exchanged for a particular item is called its "price."

FIGURE 1.2 - TYPES OF MARKETS

Markets can be physical (top) or virtual (bottom).

There are markets for all kinds of things: bicycles, furniture, gasoline, shotgun shells. Each market has a certain level of demand and a certain level of supply. Take shotgun shells, for example. In the U.S. at any given time, there are a certain number of people who want to buy them. The total amount they want to buy is called the "demand." At the same time, there are a certain number of businesses that want to sell them. The total amount they have for sale is called the "supply." The bargaining that goes on between buyers and sellers for those

WHAT IS GDP?

Gross Domestic Product: The total dollar value of what the economy produced over a given period. Also considered the nation's total income.

shells determines the price. Even though we think of prices as staying the same, they change constantly because the demand and supply is always changing.

Now, if you take all the individual markets for all the products in America and add them up, you will have one large national market. When you hear news reports about "the economy," this is usually what they are referring to. These reports use all kinds of terms when evaluating how well the economy is doing. One of the most popular is "gross domestic product," or GDP.

GDP is the total dollar value of what the economy produced over a given period (usually a quarter or year). If you added up the value of all the shotgun shells, tablets, cars, and everything else that was produced in the nation for that period, you would have the GDP. GDP is also considered the total income of the nation. The assumption is that, if you made it, you sold it, and what you sold it for is your income.

When discussing the economy, you'll also hear terms like labor, capital, unemployment, and inflation. Labor represents the workers who make things, while capital refers to the money, machines, and buildings the workers use. You need both labor and capital to produce a product. Unemployment is the percentage of people who want jobs that are not able to find them; this is called the "unemployment rate." In recent times, an unemployment rate of under 7% is generally considered good. Inflation is the amount prices rise over a year's time. A 3% inflation rate means something that cost you $100 last year will cost you $103 this year. A number of factors can cause inflation, but rates of 3-5% are considered acceptable.

FIGURE 1.3 - THE EFFECT OF DEPRESSIONS

Depressions lead many to homelessness.

Economies can experience periods of growth, known as expansions, or periods of decline, known as contractions. We use two terms when discussing periods of contractions, recession and depression. Recessions occur when the whole economy shrinks for more than a few months.[1] Recessions usually cause GDP, employment, and income to fall. Depressions,

1. William A. McEachern, *MacroEcon: Principles of Macroeconomics*, 6th ed. (Boston: Cengage Learning, 2019), 78.

on the other hand, are recessions that keep going, sometimes for years. Economists differ on how long the economy must shrink for it to be called a depression, but they generally agree that there has only been one in U.S. history, the Great Depression from 1929 to 1939. When an economy is coming out of a recession or depression, we call it a "recovery."

> **RECESSION & DEPRESSION**
>
> **Recessions** occur when the economy shrinks for more than a few months. **Depressions** are recessions that keep going, sometimes for years.

There is one final idea we need to discuss that is central to economics, and that is the concept of "rational self-interest." Rational self-interest is the assumption that the people in an economy—from consumers to producers to investors to the government—will act in a way that makes sense (rationally) and in a way that is the best for them (self-interest). Let me give you an example.

If I am deciding whether to buy a new computer, I am going to evaluate the options in an intelligent way. All other things being equal, I will conclude that a more powerful processor is better than a less powerful one (rationally, in other words). And I will buy the one that is best *for me* not for the guy I am buying it from (this is the self-interest part). This applies to everyone in the economy.

Why is this important? It is important because economics, in some sense, is really about determining how people will react in certain situations. How will people respond if the price goes up? They will buy less. How will business owners respond to an increase in demand for a product? They will make more. How will citizens respond to tax increases? They will try to avoid them (duh).

As we go through the chapters of this book, the conclusions we draw about certain actions by the different entities within the economy (consumers, producers, investors, the government, and others) will flow directly from this idea. What you will start to see is that the truths about economics are common sense and unavoidable because of how people make decisions. And, because they are common sense, they are fairly easy to understand. More about this as we move along.

THE BLESSINGS OF A GOOD ECONOMY

A good economy (one that provides well for those living under it) brings many blessings to the people of a nation and even to the world at large. A good economy allows people to get the products and services they need to live healthy, fulfilled lives. Yesterday, my wife took my son to the doctor to check on a cyst he has had in his cheek since he was born. Because of the abundance of

> **RATIONAL SELF-INTEREST**
>
> The assumption that people in an economy will act in a way that makes sense (rationally) and in a way that is the best for them (self-interest).

FIGURE 1.4 - OPTIONS FOR ECONOMIES

Economies produce blessing (top) or scarcity (bottom).

the American economy, she had no problem finding a good doctor and setting a convenient appointment date and time, and I had no problem finding the money to pay for the visit (well, at least, not much). A few minutes ago, she left to pick up her weekly grocery order at Wal-Mart. Again, because of our economic system, she could choose from thousands of products that satisfy not only our basic needs but also so many of the things we want just for the joy of consuming them (like Rocky Road ice cream).

A good economy also limits evil. When people can get jobs that allow them to support themselves and their families, they are much less likely to turn to violence and crime as ways of meeting those needs. It should be no surprise that the nations with the highest murder rates in the world are also some of the poorest (El Salvador, Honduras, and Venezuela are the top three[2]).

The United Nations Office on Drugs and Crime issued a report on the connection between poverty and criminal activity. Commenting on the report, Amber Pariona of WorldAtlas.com wrote, "Poverty and crime go hand in hand; crime drives away businesses and investors, reducing available human capital [workers] and creating an insecure environment, which, in turn, leads to more poverty."[3] In the same article, she also noted the connection between poverty and organized crime, gangs, drug trafficking, and drug and alcohol abuse.

Finally, the type of economy a nation has is linked to the type of government it has. Economic systems are tied to political systems. Most countries that have capitalistic economies have democratic governments that support political and religious freedom. Because of this, it is nearly impossible to find a capitalist

---

2. World Population Review, "Murder Rate by Country 2020." See Reference page for url.
3. Amber Pariona, "Murder Rate by Country." *WorldAtlas*, 9 January 2020.

economy where the government oppresses its people. This is not true of socialist economies.

TABLE 1.1 - BLESSINGS OF A GOOD ECONOMY
1. Provides goods and services people need.
2. Limits evil, like violence, crime, and drugs.
3. Leads to a peaceful democratic government.

Most socialist countries have hard line or totalitarian governments (the word "totalitarian" means the government has total control), and some of the greatest atrocities in human history have come at the hands of socialist governments. You may not know this, but the Soviet Union's formal name was the Union of Soviet *Socialist* Republics, and the Nazi Party of Hitler was actually the National *Socialist* German Workers' Party. I'll explain why this connection exists later. For now, just understand that there is an unbreakable connection between economies and political systems, meaning, if a nation chooses a certain type of economy, it will also likely end up with a certain type of government.

What, then, does this have to do with God? A lot, actually. As we have seen, an economy affects the ability of human beings to meet their physical needs, determines to some extent the crime and violence they endure, and leads to the type of government and freedom they will have. And because God cares about all of these things, he is very concerned about the economic systems nations use.

WHY CHRISTIANS SHOULD CARE ABOUT THE ECONOMY

For these same reasons, Christians and Christian churches should care about the economy. The type of economy we have, or the type we adopt in the future, will directly affect our lives and the lives of those we love. It can provide everyone around us with the opportunity to be happy and prosper, or it can guarantee their despair and oppression.

FIGURE 1.5 - THE NAME SAYS IT ALL

The Soviet Union was a socialist nation.

Let me make one other point about why the economy should be important to Christians. Most believers, at least in word, will confess their concern for the poor. It is not hard to find Bible verses that teach God's love for the poor and his desire that his people care for them. One of my favorite verses on this subject is Isaiah 58:10. I especially like how the King James translates it: "And if thou draw out thy soul to the

> **WHY GOD CARES...**
>
> God cares about the economy because economies affect the ability of people to meet their needs, live in safety, and enjoy religious freedom.

hungry, and satisfy the afflicted soul; then shall thy light rise in obscurity, and thy darkness be as the noon day." The phrase "draw out thy soul" has always touched me. It expresses the level of effort God expects his people to exert on behalf of the poor and oppressed of the world. If only we actually did this.

How does this connect to the economy, though? If we are to help the poor get out of poverty, and if a certain type of economic system tends to bring people out of poverty, shouldn't we, as part of our ministry to the poor, teach them how to build this kind of economy? In other words, how have we really helped the poor if we give them food but do not teach them how to efficiently produce and distribute that food through an economic system? Our love for the poor, then, should drive us to study how good economies work and the ways we can help other nations develop them. To overlook this is a dereliction of our duty to the poor and to the God who loves them.

WHERE DO WE GO FROM HERE?

I hope I have convinced you that this subject is important to God and, therefore, should be important to you. "But I can't understand economics," I hear someone say. "It's confusing." Yes, there are parts that are confusing, even for me (and I teach it). But the parts you need to understand are not confusing at all. In fact, as I said earlier, they are nothing more than common sense.

In the chapters that follow, I'll explain some of those basics to you. I'll start with a discussion of socialism and capitalism in chapter 2, including what they are and why one is better than the other. After that, I'll hit some popular economic questions facing Americans today. Should we hate the rich? Is profit really bad? Should we be concerned about the national debt? Chapters 3 through 9 address these kinds of subjects. And then, in chapters 10-12, I'll tie all of these principles together by showing you the disastrous economic situation America is headed for, the bigger reason for that situation, and how you can survive and even prosper through it.

Right now, let's talk about some very important (and divisive) terms in America—socialism and capitalism. Turn the page, and let the learning begin.

CHAPTER DISCUSSION QUESTIONS

For a better understanding of the ideas in this chapter, work through these discussion questions in a small group:

1. What does the word "economy" mean? What does it mean when we talk about a "nation's economy"?
2. Describe the importance of the words "endless" and "limited" in the definition of a national economy.
3. What is a market? Describe a market you have visited this week, whether in person or online.
4. What is "gross domestic product"? What does it have to do with the income of a whole nation?
5. What's the difference between a recession and depression? Which is worse?
6. Explain the idea of "rational self-interest" as it relates to how people make buying and selling choices.
7. What are some of the blessings of a good economy? Why are these blessings?
8. Is there a relationship between the type of economy and the type of political system a nation has? Although the chapter doesn't go into detail, why do you think a connection might exist?
9. Why does God care about the economy?
10. What reason in particular was given that Christians should care about the economy (hint: think about the poor)? Do you agree or disagree with this reasoning?

CHAPTER TWO: THE BASICS OF SOCIALISM AND CAPITALISM

The words "socialism" and "capitalism" are thrown around quite often in America today. Depending on the circle you run in, being called one or the other is either a terrible insult or a tremendous compliment. But what do they actually mean, and is there a reason to prefer one over the other?

As you may remember from chapter 1, to have a market, you need three things: consumers, producers, and prices. Producers use money, machines, buildings, and workers to make whatever it is they sell. The money, machines, and buildings are referred to as "capital," and the workers are called "labor." Consumers use the money they have (presumably from what they earned for their labor) to purchase what they need from producers. The amount paid for the items that change hands is called the "price."

The difference between socialism and capitalism isn't that one has these components and the other doesn't. The difference is in who makes the decisions about them. Let me explain.

THE FIRST GREAT WEAKNESS OF SOCIALISM

In socialism, a higher authority, usually a government department, makes these decisions. The government decides who will make what and in what amount, where things will be sold and at what prices they will be offered, and how many hours workers will work and what they will get paid for their labor. The decisions for the producing, selling, and buying of goods and services in a completely socialist economy, then, is made by the state.

> **SOCIALISM VS CAPITALISM**
>
> In **socialism**, a government department makes the decisions about products, prices, wages, and the like; in **capitalism**, or free markets, individuals do.

In capitalism, these decisions are made by individuals based on their own preferences, desires, and will. In a capitalistic economy, individuals decide what products they will make, how much they will produce, and at what price they will be offered. And other individuals, exercising their own free will, as well, decide who to buy from, how much to pay, and how many to purchase. A hallmark of capitalism, then, is freedom. For this reason, a capitalistic economy is also called a "free market" economy (in fact, the term "capitalist" was coined by a socialist making fun of free market supporters[4]).

So, which is the best type of economy? Which one most effectively and efficiently meets the needs of people? The answer comes down to who can make the best decisions about these issues—the state or the individual? Let's consider an example to help resolve this question.

Let's say a government minister is assigned to determine the number of shoes to be produced in a given year, including sizes, colors, and features. And let's say this guy is the best guy you could ever meet. He loves people and genuinely cares about the welfare of his fellow countrymen. And because of this, he wants to do the best job he can deciding all the economic factors related to shoes.

What's the problem, then? The problem isn't his motivation, or even his character; the problem is information. For him to do his job well, he will need to know what kind of shoes you will want, the exact features you will want, how many you will want to buy, and how much you will be willing to pay. He will also need to know what kind of raw materials will be available next year to make them (and which ones will not), the technology changes that will affect the number of shoes that could be produced, the cost of electricity in every factory in the country, and a thousand other things. Now, even if he spent every waking hour walking around shoe stores and factories all over the country, he could not possibly know enough to make the best decision for you and 300 million other people, could he?

Let's put ourselves in his shoes for a moment (sorry for the pun). Let's say he noticed a lot of people wearing blue and red shoes this year, so he ordered the

---

4. Kevin. D. Williamson, *The Politically Incorrect Guide to Socialism* (Washington, DC: Regnery, 2011).

FIGURE 2.1 - BAD PLANNING

Socialist planners cannot accurately predict demand.

factories to produce blue and red shoes for next year. And then tastes changed, and people preferred green and orange shoes instead. What would he do with all those blue and red shoes? Or what if he thought leather would be abundant next year, so he had all the factories gear up to produce leather shoes. But then an infection killed most of the cows in the country and the supply of leather dropped to almost zero. What would he do then?

It is this kind of information challenge that makes the socialist approach doomed from the get-go. Kevin Williamson, in his book *The Politically Incorrect Guide to Socialism,* quotes James Dorn on shoe problems in the former Soviet Union (a large socialist nation that formed after the end of World War II) due to this very issue[5]:

> The odd thing is that shortages appeared in product lines of which the Soviet Union was the largest producer in the world. In the late 1980s, the USSR produced more than three pairs of shoes for each citizen, but people had to wait to buy shoes. The problem was that the available shoes did not reflect consumers' tastes: the shoes were made to fulfill a government plan, not to satisfy market demand.

Now, consider the same example but have individual buyers and sellers in the market making all the decisions. The shoemaker uses his production and sales data from last year as a starting point for the types of shoes he will make this year. He then researches the trends in shoe preferences that are developing in the market. Based on this data, he creates a few sample shoes and brings in groups of customers to review them. He then decides to run a few hundred of each shoe to see how they sell. Based on the early sales data, and continued discussions with his customers, he makes a few alterations to color and style, and tries again. He finally gets it right and sells thousands of pairs of shoes to happy customers.

Meanwhile, you visit several local shoe stores looking for the perfect pair for an upcoming office Christmas party. You try on dozens of styles, and then pick the one that best matches your outfit and personal taste. The store where you find

5. Williamson, 39.

them is selling them for too much, you think, so you go home, find them on Amazon, and have them delivered the week before the party. Man, are you going to look good!

You can see here that both you and the shoemaker have infinitely more information than the government minister could ever hope to have. It's not that he's a bad guy, or that he's not smart enough; he just doesn't know what you know. And he is not in a position to make the necessary changes to products, prices, or preferences on a moment's notice that you and the seller are (and millions of other buyers and sellers across the country). It isn't bias that says socialism is a bad idea—it's common sense.

Now, that is not to say that free markets always get it right. There are sometimes shortages and sometimes surpluses, and some firms prosper while others fail. The question in economics, though, is not which system can be *perfect*; it is which system can be *most efficient*. In other words, which approach does the best job of using limited resources to fill the endless needs of people? On this point, free market capitalism wins by a long shot.

THE SECOND GREAT WEAKNESS OF SOCIALISM

But information is not the only weakness of a socialist economy. Another great weakness—and perhaps the most destructive to human well-being—is what socialism does to incentives. Let me explain.

The idea behind socialism (the "philosophy," if you will) is that nobody should have more than anybody else. Everybody should have the same of everything. You can see this idea taking hold in our nation when you hear people talk about "equality." To the socialist, the ultimate good is for everybody to be equal in what they have[6]; the ultimate evil is for some to have more than others.

TABLE 2.1 - WEAKNESSES OF SOCIALISM
1. Government officials lack information to make accurate economic forecasts and decisions.
2. Making everyone the same takes away the incentive to work, create, and produce wealth.
3. It leads to oppressive forms of government.

I will deal with the morality of this in a moment, but for now let us just consider what giving everybody the same thing does to both people and

6. Although, even in socialist economies, those who make the decisions for others usually have more than anyone else. The men and women who make the rules ensure the system benefits them at the expense of others.

FIGURE 2.2 - LOST INCENTIVES

Socialism eliminates the drive to work and create.

prosperity. In an ideal socialist system, everybody has the same car, the same house, the same pay, the same everything. Now, let's say you are the most talented musician ever to grace the face of the earth. Not only are you musically inclined, you are also a go-getter, dealmaker, and risk-taker. You want to hustle and make something of yourself. Your dream is to start your own music company that sells your music to the masses. You also want to help others get started in the recording business. You'll help them find success, and, when they do, you'll earn a little profit along the way. Woohoo, let's go!

There's just one problem. In a socialist economy, you can't do any of these things. You are grouped in with everybody else. No matter how smart you are, how talented you are, or how driven you are, you simply do not have the ability to use your gifts to succeed. It is not an option in the system; it is not a choice on the menu.

Now think about what that would do to your drive. How would you feel waking up every day with no opportunity to soar on the God-given talent you had? Your incentive to create that company, to provide that benefit to society, to give those other artists a chance they never could have had is gone. Now multiply this across the entire nation. All of these talented, capable human beings with no incentive to do anything. How would that affect the welfare of the people? What would it do to their standard of living, their level of income? It would destroy it, wouldn't it?

The reason America has produced some of the greatest inventions in world history, and the wealth and abundance that those inventions generate, is because people here have the incentive to do so. In America, if you use your gifts and risk your money, the payoff is unlimited if you succeed. There is no government there to stop you, and no higher power placing limits on what you can do. The world is truly yours if you are an American.

FIGURE 2.3 - INCENTIVES LEAD TO INNOVATION

Americans create great inventions because of free market incentives.

Don't believe me? Then ask yourself a question: Why was the iPhone invented here and not in Russia? Or why did the computer, video games, MRIs, cell phones, and Amazon all come from one nation—the United States of America? Because, due to the way our economy is set up, people have the greatest incentive possible to roll the dice and make their dreams come true. This kind of creativity and risk taking would never happen in a socialist system because there simply are no incentives to be creative or to take risks.

I have a good friend whose wife is Cuban-American. Every now and again, they go back to Cuba to visit her family. At lunch one day, I asked him what Cuba was like. He said, "Man, it is such a beautiful country. It's like paradise. But the people are in total despair. There are no opportunities there. Everything is so run down. No matter what they do, they can't make it better because the system won't let them."

This is an extremely important point, one I cannot overemphasize. As we move more towards socialist ideas in our country, we will slowly erode the incentives people have to take risks, create things, and bring blessings to their fellow human beings. When we beg the government to "Tax the rich!" or "Force the drug companies to lower their prices!" or "Regulate that industry—they're making too much money!" we are attacking that which has made us such a creative, productive, and prosperous people. We are, in effect, begging our government to make us like Cuba. This is why John Mackey, CEO of Whole Foods, said that "capitalism is the greatest thing humanity's ever done" and "socialism is trickle-up poverty."[7] He is absolutely right on both counts.

AN EVEN UGLIER SIDE OF SOCIALISM

In chapter 1, I discussed the relationship between economic and political systems. I noted the connection between capitalism and freedom, and socialism and oppression. The capitalism/freedom connection should be fairly obvious. If the system is set up for individuals to make free choices about what they make and buy, then the government necessarily must honor that freedom and stay out of the affairs of its citizens as much as possible. That makes sense, doesn't it?

But why did I connect socialism to oppression? This, too, makes perfect sense. Step back and consider the control the government must have to make the economic decisions we discussed earlier. For a government to decide what is made, how much is made, the prices that will be charged, and where and under

7. Stephen Sorace, "Whole Foods CEO Slams Socialism as 'Trickle-up Poverty': 'It doesn't work.'" *Fox Business*, 29 November 2020.

A CEO ON SOCIALISM

"They talk about trickle-down wealth, but socialism is trickle-up poverty. It just impoverishes everything." *John Mackey, Whole Foods CEO*

what conditions things will be sold, it must control nearly every aspect of life. By definition, it must control the factories, the land, the owners, and the laborers. It must control the truck drivers and grocery stores and power companies. Its reach would have no bounds because the economy touches every part of the nation.

What does this mean? It means that, when we ask the government to make things "fair" and "equal" by enacting socialist policies, we are asking them to take over our entire society. Ronald Reagan knew this. Let his words from more than 50 years ago ring in your ears[8]:

> A government can't control the economy without controlling people. And [America's Founding Fathers] knew when a government sets out to do that, it must use force and coercion to achieve its purpose.

Often, the road to socialism starts out with good intentions. The people get upset that a certain thing seems unfair or unjust, so they cry out for the government (or some revolutionary leader) to save them. In order to get "what's rightfully theirs," they give more and more power over to their leaders so things can eventually get "better" for them. What they find on the other side of that revolution, however, is a master far more heinous than the one they had before, but now they have given up their freedom to do anything about it.

FIGURE 2.4 - A WISE LEADER

Ronald Reagan understood that socialism meant fewer freedoms for the people.

If the Bible teaches us anything about man, it teaches that he does not do well with power. The more power you give him, the more he uses it to serve himself and harm others. In the end, human beings, regardless of their words to the contrary, always abuse their power. The following quote about Cuba's government shows how economic power eventually becomes oppressive political power[9]:

8. Williamson, 8. From "A Time for Choosing" by Ronald Reagan, delivered in 1964.
9. Ibid, 20. By Larry Solomon, *National Post*, May 2003.

> The [Cuban] government claims it takes no political prisoners. The numbers provided by human rights agencies—an estimated 500,000 since 1959, with thousands executed—tell a different story. In Castro's Cuba, it is a crime to meet to discuss the economy, to write letters to the government, to speak to international reporters, to advocate for human rights, to visit friends or relatives outside your local area of residence without government permission.
>
> ...Cubans found guilty under this criminal justice system...often serve 10 to 20 years in jail for political crimes. But most Cuban criminals are not political. A large proportion of the estimated 180,000 to 200,000 common criminals in Cuba's 500 prisons are people who broke the law by killing their own pigs, cattle and horses and selling the excess meat on the black market.

IS IT MORAL FOR EVERYBODY TO BE THE SAME?

So, what about the argument that it isn't fair for some to have more than others? Shouldn't we create a system where everybody has an equal supply of the resources of society? Isn't "equality" the better way? It actually isn't. There are several reasons for this.

First, keep in mind that God did not make us all with equal skills, talents, and abilities. The fact is, some people are smarter, more gifted, and more productive than others. This doesn't mean God does not love us and value us all the same. He does. Galatians 3:28 says, "There is neither Jew nor Greek, there is neither slave nor free, there is neither male nor female; for you are all one in Christ Jesus." But the fact that he loves us all the same does not mean he made us all equal in our abilities. He didn't.

FIGURE 2.5 - A CRUEL LEADER

Fidel Castro used socialism and tyranny to oppress and impoverish the Cuban people.

Not only that, but we are not all equally motived. Some people dream bigger than others, strive harder than others, and are willing to do things others just won't do. And some people are just lazy. They would rather sit around and do nothing than work to accomplish a goal. That's not discrimination; that's just fact.

Because of these differences, in an economic sense, some people's labor is worth more than others. All jobs are necessary and important, but not all are worth the same on the market. A heart surgeon is worth more than an auto mechanic, just as a professional football player is worth more than an elementary school teacher. And a hard-working, responsible, conscientious employee in whatever field is worth more than one who does only enough to get by, shows up late every day, and could care less about his work.

The real injustice is not in allowing some people to have more than others. The real injustice is giving those who do not deserve it the reward of those who do. *It is not* immoral to give a doctor more income than a mechanic, but *it is* immoral to force a doctor to go through years of medical school and residency to apply his immense intellect and skill and only make what a mechanic makes (with all due respect to the good work mechanics do).

Finally, let me remind you of the incentive issue. Regardless of how cruel you might find the logic of the previous paragraphs, you must face the devastating consequences of what "equality" really means—no incentive. When incentives are eliminated, poverty for all is guaranteed. There is simply no way around it.

What we should strive for is a system in which every person has equal access to opportunity, not equal distribution of rewards. We want every individual to be able to attain the highest level of success they desire, given their gifts and character. When it comes to this goal, there is not a nation on earth that gives more people more opportunity than the United States of America. This is why people from all over the world line up to get into our country (and risk their lives to cross our borders illegally). In America, women have become astronauts, poor white kids have become business owners, and a black man has become president. Great God, what a blessing it is to live in this land! Thank you so much for it.

BUT I HATE THE RICH!

America doesn't need socialism; it needs to protect the principles that made it great, those that flow from a free market, capitalistic economy. Now that we have dealt with these foundational subjects, let's move on to address some specific issues that keep some from embracing the truths of this chapter. We'll begin by talking about the rich. Should we hate them? Are they really the evil manipulators our society makes them out to be? Chapter 3 will answer these questions and more.

CHAPTER DISCUSSION QUESTIONS

For a better understanding of the ideas in this chapter, work through these discussion questions in a small group:

1. Fill in the blank for these statements. "The decisions for the producing, selling, and buying of goods and services in a socialist economy is made by the ______________. In capitalism, these decisions are made by ____________ based on their own preferences, desires, and ____________." Which of these types of systems do you prefer?
2. A capitalistic economy is also known as what kind of economy (think "free" something)?
3. Can the best-intentioned government worker make accurate decisions about what you want from day-to-day, month-to-month, or year-to-year? Why or why not?
4. In the 1980s, the Soviet Union (a socialist economy) produced more shoes than any other nation, yet their people had to stand in line to buy shoes. Why did this happen?
5. Fill in the blank: "The question in economics is not which system can be perfect; it is which system can be _______ ______________ (two words)." Why is this true?
6. Explain the "incentive" weakness of socialism mentioned in the chapter.
7. Fill in the blank: "The reason America has produced some of the greatest inventions in world history, and the wealth and abundance that those inventions generate, is because people here have the _______________ to do so." Explain why this is true.
8. How does moving towards socialist ideas (like extremely high taxes on the rich, controlling the prices that can be charged, and over-regulating industries) erode the incentives people have to create new products, inventions, and businesses? What will be the effect on the nation as we continue to do this (what will it do to our wealth as a people) and why?
9. Explain the connection between socialism and oppression. In other words, why does a socialist economy tend to make the government more controlling?
10. Is it moral for everybody to be the same (in an economic sense)? Explain.

CHAPTER THREE: THE RICH ARE NOT ALWAYS EVIL

In a 2019 article, Dan Riffle, an aide to congresswoman Alexandria Ocasio-Cortez, said, "Every billionaire is a policy failure."[10] Riffle's comment represents the feeling of a growing number of Americans. To many, the idea that a society would allow someone to earn enough to be worth $1,000 million is not only unfair but quite nearly criminal. This bias against the rich flows directly from socialist thinking. For the resources of a nation to be given equally to everyone, the rich must be demonized so their money can be taken.

FIGURE 3.1 - A FOOLISH LEADER

Politician Alexandria Ocasio-Cortez believes people should not be allowed to become rich.

In the past, the wealthy were hated because they inherited their riches from prior generations or were given it by "divine decree" (like kings and queens). However the money was acquired, those who had it kept it for themselves and used the power it gave them to help their buddies and abuse everyone else. In situations like this, it is easy to understand why people would hate the rich, fight to overthrow them, and lay claim to some of that wealth for themselves.

But is this the story of wealth in America? Not even close.

---

10. Roxanne Roberts, "Why does everybody suddenly hate billionaires? Because they've made it easy." *The Washington Post*, 13 March 2019.

HOW THE RICH GET RICH IN AMERICA

Yes, there are a lot of "rich" people in America, but how did most of them come into that wealth? Have some of them inherited it? Sure. Some kids grow up with a silver spoon in their mouths and have everything handed to them. It's a fact, but it's not a crime. And the reality that we don't like it doesn't give us the right to hate them or petition the government to take away that wealth.

WEALTH CREATION USA

In free markets like America, the vast majority of wealthy people earn their wealth by hard work, risk-taking, and value creation, not inheriting or stealing it.

In America, the vast majority of wealthy people *earn* their wealth. They start off as regular Joes like you and me, work hard, take risks, and bring a great deal of value to the market. It is that value—not greed or bribes or theft—that eventually leads to their wealth. This is a critical point to understand: Wealth in a free market comes from value creation.[11]

Just what is value creation? Value is something others are willing to pay for. For the last few years, I have been investing in my education. I have spent thousands of dollars of my own money on an MBA, plus additional courses in marketing and economics. I have done this to increase my value. Colleges need teachers, and the more subjects you can teach, the more valuable you are. I am even finishing a doctorate in business to make my value even greater. In a sense, I am creating value that I want to turn into income in the future.

Now, I do not expect to become wealthy teaching. The value I create will be enough to earn a good living but not enough to create true wealth. The value that some people create, however, is enough to generate incredible riches.

Take Larry Ellison, for example. Larry founded a firm called Oracle in 1977. If you've ever worked for a big company, you are probably familiar with Oracle software. It isn't the kind of software you put on your home PC, though; it's the expensive kind

FIGURE 3.2 - VALUE CREATORS

Larry Ellison created Oracle software, a product so valuable it made him wealthy.

11. Obviously, some people use that system to illegally gain wealth. There are con artists, liars, and criminals who become wealthy in America. But the vast majority of wealthy people build their wealth legally by creating value the market desires.

used to run the world's biggest companies. It helps them process transactions, manage their customers, and store and use the data they collect.

Okay, so maybe you haven't heard of it. Well, all you really need to know is that it's very successful. I mean, *really* successful. According to Forbes magazine, the company is so successful that Larry, Oracle's founder, is now the fifth richest man in the world, with a net worth of $59 billion.[12]

But how does starting a successful company lead to wealth? Here's how it works in very simple terms. Larry is the creator of this valuable product, Oracle software. Now, even though he doesn't do all the work at Oracle, he is the legal owner of the company (and the reason for its existence), so, every time a customer recognizes the value of Oracle software and buys a copy, Larry gets a little of the money from that sale.[13] It is his payment for the value he has created. The more people recognize and pay for the value of Larry's product, the more money he gets. And, at some point, as the product becomes more and more successful, Larry becomes wealthy from all the money he has made.

There are a couple of important points to take from this example. First, Larry did not force anyone to buy his product. They freely chose to buy it because of its value. They wanted it, so they bought it. This is true of men like Jeff Bezos, the founder of Amazon, too. The next time you order something on your Amazon Prime account, ask yourself, "Did Jeff make me do this?" The answer, obviously, is no. You freely gave Jeff some of your money because of the value he has brought to you.

Now, here's a question to ponder: Why do we hate the men that bless us with so much value? We hate them, which implies we want them destroyed, but if we destroyed them, we would lose the value we are so glad they brought us. Are we insane or what?

Think about how we treat Wal-Mart these days. Many seem dedicated to putting it out of business for no better reason than it is "big and rich." Among other things, Wal-Mart is accused of underpaying workers (though none of their employees are forced to work at Wal-Mart) and causing the little mom-and-pop shops to close (though Wal-Mart didn't do this; the customers did by buying from Wal-Mart instead of the mom-and-pops).

But how did Wal-Mart get so big? Did Sam Walton, the founder of Wal-Mart, go around and put a gun to his customers' heads and force them to buy

12. Forbes.com, "The Richest in 2020."
13. Larry isn't the "owner" of Oracle anymore. It is a publicly traded company on the NASDAQ. However, the illustration still holds.

from him? No, he sold quality products at such cheap prices that people came in droves to buy from him. Have you ever been to a Wal-Mart on a Saturday morning? You have to beat the customers off with a stick. Why? Because Wal-Mart has singlehandedly reduced the prices of almost every retail product on the market. In fact, most of us could not live the kinds of lives we live without Wal-Mart.

FIGURE 3.3 - DEMONIZED COMPANIES

Wal-Mart is attacked because it is successful.

And yet we hate them. Why?

The second point to take from this illustration is about incentives. If you take away the incentive for Larry Ellison, or Jeff Bezos, or Sam Walton to take the risks they took to create these fantastic companies, would any of them ever have been created? If all Larry could get from starting Oracle was the same car, pay, and home he had before he started it, why on earth would he go to all the trouble? He wouldn't.

What does that mean, then? It means that if you want socialism in America, you can say goodbye to Oracle and Amazon and Wal-Mart and every other company that brings you the value that makes your life better. This is why I said in the previous chapter that socialism leads to poverty.

Before I leave this section, let me address one other reason people often give to justify their hatred of the rich, the claim that they don't pay their fair share of taxes. I will deal with this in detail in chapter 4, but for now, let me just point out that the top 1% of wage earners in America (the richest people in the nation) pay more tax than the bottom 90% combined. Yes, you read that right—the richest 1% pay more tax in total dollars than the bottom 90%. The facts simply do not support the oft-repeated mantra that the rich do not pay their fair share of taxes.

WHY DO WE REALLY HATE THE RICH?

Now, back to why we hate those who bring us so much value. When you understand that rich people in America have earned their wealth because of the value they have brought to consumers, and that people freely choose to give them money in exchange for that value, you start to realize there must be other motives for our hatred. What are they?

THE RICH & TAXES

Contrary to popular belief, the rich in America pay the vast majority of income taxes. In fact, the top 1% pay more tax that the bottom 90% combined.

TABLE 3.1 - WHY WE HATE THE RICH
1. We are jealous of their success and lifestyles.
2. We are greedy and want what they have earned.
3. We are told to hate them by our leaders.

In the first place, it is jealousy, pure and simple. They have a lot, we don't, and we don't like it. It is human nature to look down your nose at the guy driving the big BMW, wearing the $1,000 suit, and spending most of his weekends in Cancun. We look at the Larry Ellison's of the world and burn with envy. We are jealous of their intelligence, their drive, and their creativity. And we are jealous of their success. So we hate them.

Our hatred is also driven by greed. It's ironic, isn't it? Every day, millions of Americans accuse the rich of being greedy and materialistic, but if they were to look into their own hearts, they would see that that is the very reason they hate them—because *they* are greedy and materialistic themselves. They want what these other men and women have worked for; they just don't want to work as hard as they did or risk what they risked or endure what they endured to have it. Remember, you don't have to be rich to be greedy.

There is one other reason we hate them—because we are told to. There are certain groups in our country who gain power by making us hate others. If they can make you a victim of the "rich, white guys," they can get your votes. And, if they can make enough of you hate them, they can claim power over the entire country. And they are doing a fantastic job of it.

FIGURE 3.4 - RICH HATERS OF THE RICH

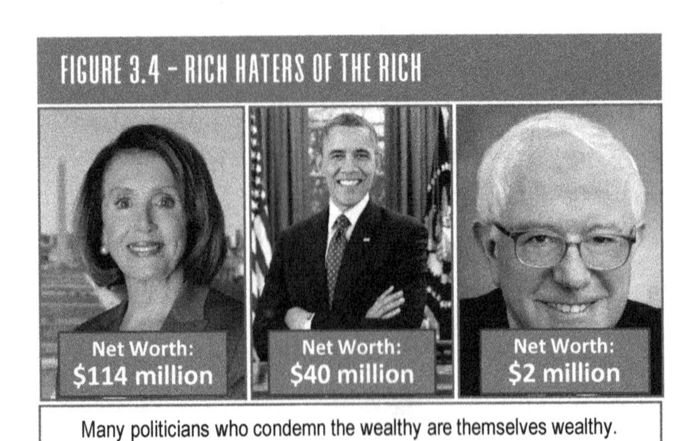

Many politicians who condemn the wealthy are themselves wealthy. (Left to right: Nancy Pelosi, Barack Obama, and Bernie Sanders)

Ironically, most of the people who tell you to hate the rich are rich themselves. Nancy Pelosi, the Democrat senator from California, is worth $114 million[14]; Barack Obama, former president of the United States, is worth $40 million[15]; and Bernie Sanders, former presidential candidate, is worth $2

14. Megan Henney, "How much money is Nancy Pelosi worth?" *Fox Business*, 17 July 2020.
15. Margaret Abrams, "Barack and Michelle Obama net worth 2020: How much is the former US President worth along with his wife?" *Evening Standard*, 19 February 2020.
16. Chole Foussianes, "How Bernie Sanders Became a Millionaire." *Town and Country Magazine*, 15 March 2020.

million.[16] Each of these politicians has demonized the wealthy at various times for political purposes. Why would they do this? Why condemn the very thing they are? Because it works. We are foolish enough to keep falling for it, so they keep doing it.

WHAT GOD CAN TEACH US ABOUT HOW TO VIEW THE RICH

Does God hate the rich, too? I've heard politicians and preachers alike condemn wealthy people using the Bible. A popular place to go is James 5:1-6. Let me quote it for you here.

> Come now, you rich, weep and howl for your miseries that are coming upon you! [2]Your riches are corrupted, and your garments are moth-eaten. [3]Your gold and silver are corroded, and their corrosion will be a witness against you and will eat your flesh like fire. You have heaped up treasure in the last days. [4]Indeed the wages of the laborers who mowed your fields, which you kept back by fraud, cry out; and the cries of the reapers have reached the ears of the Lord of Sabaoth. [5]You have lived on the earth in pleasure and luxury; you have fattened your hearts as in a day of slaughter. [6]You have condemned, you have murdered the just; he does not resist you.

These are some pretty strong words against the rich. Maybe God doesn't like them after all? Not so fast. Let's take a closer look at the context to see what is really going on here.

Why is James, and, by extension, God, so upset with the rich in these verses? *Because they used their power to oppress the poor.* Verse 4 says they did not pay their workers what they were owed, and verse 6 says they had unjustly condemned the needy and even had them murdered. All the while, they were living large on the ill-gotten gain they had stolen from them.

Now, I ask you, do the rich in America do this? Sure, from time to time, some rich guy might not pay his workers what they are owed, and, certainly, there have been rich folk involved in murder. But what happens to them when they do these things? *They go to jail.* Jeff Bezos doesn't just refuse to pay his employees so he can take another trip to Tahiti. If he did, he would be arrested, and rightly so.

And don't forget, there were many righteous rich people in the Bible. Abraham was wealthy (Genesis 13:2), Job was wealthy (Job 1:3), Boaz was wealthy (Ruth 2:1), and Joseph of Arimathea was wealthy (Matthew 27:57). God is not

FIGURE 3.5 - WEALTHY BIBLE FIGURES

Abraham and other men in the Bible were wealthy.

against the wealthy any more than he is against the poor. He is against wickedness, no matter what the financial condition of the person committing that wickedness may be. You can be rich and be a godly man, or you can be poor and be a wicked one. Wealth is irrelevant to God.

BUT A LOT OF RICH PEOPLE ARE GREEDY SCUMBAGS

Yes, some rich people are greedy, unrighteous, godless people. No doubt about it. So, let me ask you: What's your solution? Enact socialism and make everybody poor because that rich guy is such a pompous jerk? Where will that leave us?

What you need to realize is that a free market economy is the only economic system that turns that rich guy's greed and arrogance to your favor. You see, because of the law of incentives, which we discussed at length earlier, that greedy jerk gets up every morning and goes out and tries to bring value *to you*. He wants to make that product just like you want it so you will buy it and he can pile up more money in his bank account. And that arrogant, rich doctor can't wait to perfect that new surgical tool to save thousands of lives—like yours—so he can get his picture on the cover of a magazine.

Sin and sinful people will exist in either economic system, socialism or capitalism. The beauty of capitalism is that it uses the sin of man to bring blessings to mankind. Socialism, on the other hand, just destroys everybody, rich and poor, righteous and sinner alike.

Don't get me wrong. The Bible condemns both arrogance and greed. Proverbs 16:5 says, "Everyone proud in heart is an abomination to the Lord; though they join forces, none will go unpunished." God hates pride, and he has promised to punish it, in this life and the next. He also stands against greed and promises that those who practice it will come to ruin. Proverbs 15:27 reads, "He who is greedy for gain troubles his own house." But the answer to these tendencies in mankind is not to place the entire race under a system that leads everybody to poverty. The answer is to let them live

THE BEAUTY OF CAPITALISM

Because of the law of incentives, a free market economy is the only system that uses greed and arrogance to bring blessings to other people.

under a system that will give them all the opportunity to prosper, while also teaching them the dangers prosperity brings.

BUT THE POOR!

So far, we've only talked about people who have a lot of money. But there are still poor people in free markets. Doesn't socialism do more for them than capitalism? Don't they deserve more than they get? Let's save that discussion for the next chapter. Turn the page and consider the reality of the poor in America.

CHAPTER DISCUSSION QUESTIONS

For a better understanding of the ideas in this chapter, work through these discussion questions in a small group:

1. How do most Americans become rich? What does the idea of "value creation" have to do with it?
2. Do men like Jeff Bezos, the founder of Amazon, force people to give him their money? If not, why do people give men like Jeff so much?
3. The chapter stated the following: "We hate [the rich], which implies we want them destroyed, but, if we destroyed them, we would lose the value we are so glad that they brought us." Does this truth about our hatred of the rich make any sense? Why or why not?
4. What would happen if we took away the incentive from men like Larry Ellison (Oracle), Jeff Bezos (Amazon), and Sam Walton (Wal-Mart) by preventing them from becoming rich off the companies they started? Would these big companies still be created if their founders could not earn wealth? What impact would that have on the people of our country?
5. Fill in the blank: "The top 1% of wage earners in America (the richest people in the nation) pay more tax than the bottom ____________% [a number between 1 and 100] combined." Did this fact surprise you? Why or why not?
6. Name some reasons the chapter suggests that we really hate the rich. Which of these do you think applies the most?
7. How much are the following people worth: Nancy Pelosi: $ _________ million. Barack Obama: $ _________ million. Bernie Sanders: $ _________ million.
8. Based on James 5:1-6, when does God condemn the rich? What do they do in these verses that he hates?
9. Name some rich men from the Bible. Were you surprised so many righteous men were wealthy?
10. How does a free market economy use the greed and arrogance of a rich man or woman to bring blessings to everyone?

CHAPTER FOUR: THE POOR ARE NOT ALWAYS GOOD

In America today, there is an awful lot of talk about the poor. No matter what economic issue the nation is facing, somehow the poor always get dragged into the conversation. If you believe the arguments of some politicians, the poor here are worse off than anywhere else in the world. In fact, in spite of the stellar work ethic and character of the needy, wealthy individuals and corporations oppress them mercilessly. The only solution, then, is to adopt socialist policies that will force the rich to give the poor what is rightfully theirs. But is this the correct approach? And, more importantly, is this the true plight of the poor in this nation? Let's take a closer look and see.

THE REAL STATUS OF THE POOR IN AMERICA

We'll consider this question from two perspectives. Firstly, let's view it from the perspective of the wages paid to low-income workers. Do rich companies actually exploit them by paying them less than they deserve? And secondly, let's examine it from the perspective of taxes and welfare assistance. Do rich individuals skip out on paying taxes, thereby denying the poor the benefits they need?

Regarding the issue of unfairness in wages, let me first remind you of the rules for employment in America. Given the laws of this country, can workers be forced to stay at an employer they don't want to work for? No, of course not. They can quit at any time and take another job. Also, can one employer, to gain the services of an employee working for someone else, offer to pay that employee more than their current employer is paying them? Sure, they can. If I think you would make a valuable addition to my staff, I can offer to pay you whatever I want. This is the way it works in America (this is why we call it a *free* market).

These characteristics of the employment market—that workers can move freely between employers, and that employers are in competition for workers—makes it impossible for companies to take advantage of workers in a wholesale fashion. Consider the example of Wal-Mart. Suppose Wal-Mart in Any City, USA, is paying its workers too little for their skill and capabilities. What would the employees do when they realized this? They would look for jobs at employers who would pay them what they were worth, wouldn't they? Of course, they would.

Now, let's say there is a Target directly across the parking lot from Wal-Mart, and this Target needs workers. Now, if Target knew Wal-Mart was underpaying good employees, what would it do? Advertise its jobs at more competitive hourly wages. And then what would happen? Over time, workers would move from Wal-Mart to Target at higher wages. In response, Wal-Mart would either go without workers (which is highly doubtful) or offer higher wages to keep the employees it had and convince others to come back. This competitive dynamic prevents individual employers from forcing employees to accept wages below their value.

So, how do we explain the fact that many employees feel like they are being paid too little? *Feeling* like you are underpaid and *actually being* underpaid are two different things. In a free market, the buyers in the market decide what something is worth. In the case of the value of an employee, the buyers are the employers.

In a sense, when you look for a job, you are selling your labor on the market, and Wal-Mart and Target and Lowes and Apple are the buyers. Employers pay employees based on their actual value to the company, not what those workers think they are worth. So, if you apply for 10 similar jobs at 10 companies and get offers of $12-$14 per

FIGURE 4.1 - COMPETITION CREATES FAIR WAGES

Competition between companies for labor makes it impossible to pay workers less than their worth. (Target image by Farrgutful.)

hour, that is the market rate for your skill, experience, and capabilities in that job. You cannot change that just because you don't like it.

The fact is, if you are offered $12-$14 per hour, that is all you are worth to the market. That does not mean you, as a person created in God's image, are only worth that; it means you, as a worker in a competitive market, are only worth that. This is the cold, hard truth that many overlook in today's "equality" society.

The answer to feeling your wages are too low, then, is to make yourself more valuable, not force your employer to pay you a higher wage because you want him to (more about the problems with minimum wage laws in chapter 6). There are tons of jobs out there that pay way more than minimum wage, but you have to work to become qualified for those jobs. This comes through investing in yourself via training, education, and experience. You, then, are the answer to your low wage problem, not your employer or the government.

Now, let's examine the question of the treatment of the poor in America from the perspective of income taxes and welfare assistance. Do the rich pass the burden of taxes off to the poor, keeping the funds that could have been used to help them tucked away in their greedy pockets? According to the Tax Foundation, the wealthy pay the vast majority of income taxes in the U.S.[17] In 2017, for example, the top 1% of America's wage earners, those with income of $515,000 or more, paid $616 billion in income taxes. The bottom 90%, however, paid only $479 billion. That means that the top 1% paid more income tax than the bottom 90% combined (28% more, in fact).

As a share of all income taxes paid, the top 1% paid 39%, the top 10% paid 70%, and the top 25% paid 86%. By the time you get to the top 50%, those with income of around $42,000 or more, you have accounted for nearly all the income taxes paid in the U.S., 97%. To put it another way, the bottom 50% of wage earners (those making less than $42,000 in adjusted gross income) paid only 3% of all income taxes.

In terms of the percentage of income that goes to taxes, the burden is equally skewed toward the rich. For example, the bottom 50% of wage earners paid an average tax rate of only 4%, meaning, for every $100 they earn, they paid only $4 in taxes. The top 1%, on the other hand, paid an average rate of nearly 27%, or $27 per $100 in earnings, almost seven times as much. Table 4.1 summarizes the taxes paid by wage earners in the U.S.

What does all this mean? It means the claim that the rich do not pay their

17. Erica York, "Summary of the Latest Federal Income Tax Data, 2020 Update." *Tax Foundation*, 25 February 2020.

fair share of taxes is not true. In fact, they pay a great deal more in both total dollars and as a percentage of income. If you believe it is fair that the rich pay more taxes than the poor, then you should be well-satisfied with the U.S. tax system.

TABLE 4.1 - TAXES PAID BY AMERICAN TAXPAYERS

Tax Payer Category	Income Level (and up)	% of Total Tax Paid	Average Tax Rate
Top 1%	$515,000	39%	27%
Top 10%	$145,000	70%	24%
Top 25%	$84,000	86%	18%
Top 50%	$42,000	97%	16%
Bottom 50%	$42,000	3%	4%

So, what about welfare assistance? Do the rich prevent the poor from receiving the benefits they need? In 2013, the Cato Institute performed a state-by-state study to determine if a person was better off working a minimum wage job or staying on welfare (from a financial perspective).[18] The study found that, at that time, welfare paid more than a minimum wage job in 35 states, and, in 13 states, welfare benefits exceeded $16 per hour in 2019 dollars. Hawaii had the highest value of benefits at almost $54,000 per year, the District of Columbia was second at slightly more than $47,000, and Massachusetts third at just under $47,000. In hourly rate terms, this works out to between $23 and $26 per hour.[19] The bottom three were Arkansas, Tennessee, and Mississippi, all hovering around $19,000 per year. Even at these lower benefit levels, the hourly wage still works out to more than $9 per hour.

To put this in perspective, an article by Forbes Magazine noted that if you had an annual income of $9,000 or more per year in 2015, you would fall into the top 20% in the world.[20] Now, that is not to say someone would want to live on $9,000 a year in the U.S., but it does help show the generous level of benefits the poor receive in this country. Table 4.2 gives the benefit values and equivalent hourly wage rates for select states around the country.

---

18. I was unable to find a comprehensive study any later than 2013. Numbers reported here are adjusted by inflation through 2019, the last year inflation numbers were available at the time of this writing. Michael Tanner and Charles Hughes. "The Work vs. Welfare Trade-Off: An Analysis of the Total Level of Welfare Benefits by State." *Cato Institute*, 2013.
19. To get these hourly wages, divide the yearly value of the benefits by 2,080 (a 40-hour work week times 52 weeks per year).
20. Tim Worstall, "The Average US Welfare Payment Puts You In The Top 20% Of All Income Earners." *Forbes*, 4 May 2015.

TABLE 4.2 - WELFARE BENEFITS BY STATE

State	Total $ Value (2019 dollars)	Hourly Rate
Hawaii	$54,000	$26
Washington D.C.	$47,000	$23
California	$39,000	$19
Washington	$34,000	$16
North Carolina	$31,000	$15
Virginia	$23,000	$11
Texas	$20,000	$10
Mississippi	$19,000	$9

Not only are we generous to our poor citizens, we are also generous to our poor immigrants (legal and illegal). A Cato Institute study of the welfare benefits paid to noncitizen immigrants, including those with green card status, refugees, asylees, temporary migrants, guest workers, and illegal immigrants, found that in 2016 the average noncitizen immigrant received almost $4,000 in welfare benefits compared to just over $6,000 paid to the average citizen.[21] This number is found by dividing the total amount spent on immigrants by the total number of immigrants, whether or not they actually qualified for welfare. The value of benefits paid to those who actually qualified was around $16,000 per immigrant versus $22,000 for each citizen.

What do we conclude, then, about the relationship between the rich and poor in the U.S.? Do the rich really oppress the poor? Not in the slightest. In fact, far from *oppressing the poor,* the rich actually *support the poor* in America. Without the rich, the poor would have nothing.

You can see this by asking where the money for these welfare programs comes from. As we saw earlier, it comes from the taxes paid by the wealthy. Remember, almost 97% of the income taxes paid in America are paid by the top 50% of wage earners (as well as 86% by the top 25% and 70% by the top 10%). If it wasn't for the wealthy, then, the poor in America would receive no support, because there would be no money to give them.

FACT: THE RICH SUPPORT THE POOR

Due to the high taxes on the rich, low taxes on the poor, and generous welfare benefits, the rich do not oppress the poor in America—they support them.

INCOME INEQUALITY

While we are on the subject of the poor, let me take a moment to address the often-raised complaint of "income inequality" in America. People who make this

21. Alex Nowrasteh and Robert Orr, "Immigration and the Welfare State: Immigrant and Native Use Rates and Benefit Levels for Means-Tested Welfare and Entitlement Programs." *Cato Institute*, 10 May 2018.

charge justify it by comparing the income of the richest people in the nation to that of the poorest. As time goes by, the super-rich get richer, but the poor stay poor. This gap, then, between the rich and poor is viewed as immoral. Is this a legitimate argument? It is not, in fact. There are several reasons for this.

In the first place, as we just learned, the poor in America are in the top 20% of income earners in the world. This makes it hard to justify complaints about how much rich people make when the poor are being provided for so well by those same rich people. If I could talk directly to those on welfare concerned with "income inequality," I would ask them why they even care. They aren't working, and that rich guy (who is working, by the way) is giving them enough to live in the top 20% in the world. Why are they complaining?

Secondly, as also pointed out earlier, the rich in America don't get rich by oppressing the poor; they get rich by creating value that improves peoples' lives. So, whatever is causing the poor to stay poor, it isn't the rich guy. He's not oppressing the poor to get his additional wealth, so why is he being blamed for "income inequality" when he's got nothing to do with it?

And, finally, consider this: you would expect the rich to get richer in an advanced, free economy such as ours with no limits on earning power. In fact, we should be worried *if people aren't getting wealthier* at the top levels of income within our nation, because this would mean value wasn't being created and the economy had leveled off or was declining. Remember from chapter 2, our goal should not be that everybody has the same of everything, but that everybody has the same access to opportunity, which they do in our nation.

This should be a great encouragement to the poor. All they have to do is invest in themselves, work hard, and take risks. If they do, with all the opportunity in America, they can take part in the abundance this great nation produces without having to have someone else give it to them. The charge of "income inequality" in the end, then, is just another way to make people feel oppressed so they can be controlled by those who want power over them.

THE BIBLE'S DEFINITION OF THE POOR

The U.S. government spends a tremendous amount of money to help the poor. But is all of this spending justified? Is everyone receiving benefits really "poor" in the sense that they can't do any better on their own? Furthermore, are we bound by God to help every person with a low income, or are there other factors to consider when extending our hands to the needy? Let's consult the Bible to find out what kind of poor people we should be helping and what kind we

FIGURE 4.2 - A GOD OF COMPASSION

God required part of the harvest be left for the poor.

should not.

The Bible talks a lot about the poor. From the founding days of Israel, God instructed his people to care for the needy. Deuteronomy 15:7 through 8 says, "If there is among you a poor man of your brethren, within any of the gates in your land which the LORD your God is giving you, you shall not harden your heart nor shut your hand from your poor brother, but you shall open your hand wide to him and willingly lend him sufficient for his need, whatever he needs." In fact, to provide something for the poor during the time of harvest, God required his people to leave some of their produce in the field. "When you reap your harvest in your field, and forget a sheaf in the field," the Creator said, "you shall not go back to get it; it shall be for the stranger, the fatherless, and the widow, that the LORD your God may bless you in all the work of your hands" (Deuteronomy 24:19). The Lord Jesus reflected his Father's compassion for the poor. "But when you give a feast," he told his followers, "invite the poor, the maimed, the lame, the blind. And you will be blessed, because they cannot repay you; for you shall be repaid at the resurrection of the just" (Luke 14:13-14).

Though the Bible mentions the poor throughout its pages, it is important to note the distinction it draws between the poor and the lazy. Proverbs 20:13, for example, warns, "Do not love sleep, lest you come to poverty; open your eyes, and you will be satisfied with bread." Note that the person mentioned in this verse lives in poverty, but the cause of his poverty is not oppression by the rich—it is his own laziness. Instead of working, he chooses to lay in bed all day. Proverbs 19:15 expresses a similar truth: "Laziness casts one into a deep sleep, and an idle person will suffer hunger."

What does the Lord expect us to do for this person? Nothing. He must endure the consequences of his laziness so that he will learn that work is the way out of poverty, not charity. Proverbs 21:25 affirms this brutal truth: "The desire of the lazy man kills him, for his hands refuse to labor." Indeed, God has great compassion for those who are truly poor, but he has very little for those who are lazy. In fact, his command is clear: "If anyone will not work, neither shall he eat" (2 Thessalonians 3:10).

This distinction is important for us to understand in modern America. Not everyone who falls below the poverty line is there because of injustice. Many are there because they have chosen to work the system instead of working a job. Instead of investing in themselves through education, and taking a job where they can contribute to society and earn their own way, they have chosen to sit around, do nothing, and take from those who work. This is a disgusting thing in the eyes of the Creator.

FIGURE 4.3 - WORK IS GOOD IN GOD'S EYES

God requires people to work to provide for themselves.

I have personally seen this kind of laziness and lack of integrity in action. Back in the early 2000s, the church I pastored was helping a woman named Susan. Susan was a poor single mother with two sons (I was actually the Big Brother to her oldest son). She had both of her boys out of wedlock by a man who was then in prison. Most of her income was from various forms of welfare assistance, although she did work about 20 hours a week at a minimum wage job.

After about a year of helping Susan pay her electric bill, buy groceries, and provide clothes for her children (among other things), I sat down and had a talk with her about how she could eventually get off welfare and provide a better life for her family. What I had seen was that the welfare programs were keeping her in poverty, not helping her get out of it. I told her she needed to work 40 hours a week so she could begin to transition out of welfare. The church even offered the help her make that transition. She refused.

Interestingly, Susan would walk miles early in the morning to stand in line to get another government benefit, but she would not work more hours at a job that would give her more power over her life. She did not have the integrity or work ethic to lift herself out of poverty. To my knowledge, she is still on welfare today.

I had a similar experience in 2012. I was pastoring a church in Andrews, SC, and had started a handyman business to make ends meet. One day, I was doing a bathroom remodel at a trailer in town. A man living next door caught my attention and asked if I needed any help. I did, so we worked out a deal where I would pay him hourly and he would help me with my projects.

This man was very skilled. He could do plumbing, carpentry, electrical work—just about anything. On one job, we wormed around under the crawl space of a house for two days to put in drain lines for the sinks and toilets. There was virtually nothing he couldn't do.

One day he called me and said, happily, "Hey, Wes, I can't work with you anymore. I just got my disability!" I was in shock. "What for?" I asked. Apparently, he had applied for disability for a "shoulder injury." It had finally come through, and he was getting back pay of over $20,000 plus about $1,200 a month from then on. He then informed me, "Yeah, if you want me to do any more work, I'll need you to pay me under the table." I told him I could not do that, and we never worked together again.

The thing is, he worked with me for months, did all kinds of jobs, and never complained once about his shoulder. Like Susan, he was working the system. He was of such low character and integrity that he would defraud the government of assistance so he could sit around and do nothing.

Why do I mention these examples? To show you that, just because you are "poor," or even on "disability," doesn't mean you are actually needy. Many people today—and the number seems to grow by the minute—are lazy and unrighteous and are no more deserving of the benefits they receive than crooks or con artists.

We are quick to vilify the rich in this nation, but the poor are sometimes more wicked than any rich man. In light of this reality, the comments made by the Cato Institute report mentioned earlier seem appropriate:

> The current welfare system provides such a high level of benefits that it acts as a disincentive for work. ... If Congress and state legislatures are serious about reducing welfare dependence and rewarding work, they should consider strengthening welfare work requirements, removing exemptions, and narrowing the definition of work. Moreover, states should consider ways to shrink the gap between the value of welfare and work by reducing current benefit levels and tightening eligibility requirements.[22]

THE BREAKDOWN OF THE FAMILY AND THE INCREASE OF POVERTY

Sometimes I think the accusation that the rich oppress the poor in America is a smoke screen designed to keep people from looking at the real culprit. If you analyze the reasons underlying the generational poverty of many in the U.S., especially in minority communities, one of the primary culprits you will find is the

---

22. Tanner and Hughes, "The Work vs. Welfare Trade-Off."

breakdown of the Bible's definition of the family.

As far back as 1999, data was amassing demonstrating the link between family structure and poverty. That year, the Heritage Foundation published a research report entitled, "How Broken Families Rob Children of Their Chances for Future Prosperity."[23] According to this document, "almost half of American families experience poverty following a divorce, and 75 percent of all women who apply for welfare benefits do so because of a disrupted marriage or a disrupted relationship in which they live with a male outside of marriage." The report went on to connect broken homes to other negative effects on children, including poor physical and mental health, increased sexual activity at younger ages, higher likelihood of giving birth outside of marriage, and lower educational achievement.

FIGURE 4.4 - THE HOME MATTERS

The loss of a faithful mother and father in the home is linked to the increase of poverty, crime, and violence.

On the issue of education, a 2015 report in *Education Next Journal* discussed the relationship between single parent families and the academic achievement of the children, a key factor in determining income levels.[24] The report found that, from 1960 to 2013, the percentage of black children living with only one parent rose from 22% to 55%, a two-and-a-half-fold increase. The percentage of white children living with one parent more than tripled during that period, from 7% to 22%.

Regardless of race, the college completion rate of children living with a single parent at any time between the ages of 14 and 16 was just over 13%, while those living with both parents during that age span was nearly 40%. The report concludes, "Children who live with a single parent between the ages of 14 and 16 have consistently completed less schooling than children from two-parent families, with the gap widening over time." Lower educational levels mean less earning power and a greater chance of living in poverty.

This research supports the notion that God has designed a child to need

---

23. Patrick Fagan, "How Broken Families Rob Children of Their Chances for Future Prosperity." *The Heritage Foundation*, 11 June 1999.
24. Kathleen M. Ziol-Guest, Greg J. Duncan, and Ariel Kalil, "One-Parent Students Leave School Earlier." *Education Next*, Spring 2015.

both a loving father and mother at home to grow into a productive, prosperous adult. Children need discipline, guidance, and support from their parents to develop their skills and demonstrate value to the market. But our society has dismantled God's design block by block over the last 75 years, and the result is nothing short of catastrophic.

So, what is God's design? It is actually very simple. A man and woman get married for life, making a public commitment to one another before God. The man works outside the home to earn the income needed for his wife and children, and he provides love and leadership to his family. The woman works at home to manage the house and provide the nurturing the children need to grow into healthy adults. She also supports her husband in his efforts to carry out his role as provider. The children, then, are educated in a godly, Bible-centered way, either at home by their mother or in a Christian school by other believers. In the vast majority of cases, when this plan is followed, productive, honest, well-balanced citizens are the result, the very kind our nation so desperately needs.

If we want fewer poor among our citizens, we must stop ignoring the Creator's design for the family. On the contrary, we must make restoring it a central part of our strategy. As long as we ignore this key factor, the problem of the poor, and the financial burden of supporting them, will continue to plague this nation.

BUT PROFIT—IT'S SO EVIL!

In spite of socialist arguments to the contrary, the rich are not always evil, and the poor are not always good. But what about profit? Isn't making money off the needs of others an abuse of our fellow human beings? This common claim is next on our list. So, what about profit? Should it be outlawed too? Answers are a mere page away.

CHAPTER DISCUSSION QUESTIONS

For a better understanding of the ideas in this chapter, work through these discussion questions in a small group:

1. Explain the rules that exist in America that tend to prevent rich businesses from taking advantage of poor workers.
2. Explain how competition between employers (like Wal-Mart and Target) keeps them from paying their employees less than they are truly worth.
3. Fill in the blank: "___________ are the answer to your low wage problem, not your employer or the govern-

ment." Why is this true?

4. According to the chapter, in 2017, the top 1% of U.S. wage earners paid $ __________ billion in income taxes, while the bottom 90% paid $ __________. So, who paid more? Does this surprise you? Why or why not?
5. According to the chapter, the top 50% of wage earners paid what percentage of the income taxes in the nation: 50%, 75%, or 97%? Again, were you surprised? Why or why not?
6. The chapter stated, "Welfare paid more than a minimum wage job in 35 states, and, in 13 states, welfare benefits exceeded $16 per hour. Hawaii had the highest value of welfare benefits at just over $54,000 per year, the District of Columbia was second at $47,000, and Massachusetts third at just under $47,000. In hourly rate terms, this works out to between $23 and $26 per hour." What does this mean for the claim that the poor in America are not provided for by the rich?
7. Is the following statement true or false? "If you are on welfare in America, you are in the top 20% of income earners in the world." Explain your answer.
8. Share some Bible verses from the chapter that show God's love for the poor. Which one moved you the most?
9. Explain this statement: A man who is poor because he is lazy is not poor in God's eyes.
10. How do such high levels of government benefits for the poor actually keep them from working? Explain your answer.

CHAPTER FIVE: PROFIT IS WHY YOU HAVE WHAT YOU HAVE

If there is one thing Americans seem to hate more than the rich, it is profit. Because of the advance of socialistic ideas, a goal that was once taken for granted has now become a dirty word. Anyone who seeks it must do so silently, and anyone who finds himself with too much of it must run and hide. How does this view of profit stack up, though? Is it unnecessary for prosperity, or, worse yet, it is a sin against God? Let's find out.

WHAT IS PROFIT?

We'll begin defining profit by talking about what it is not. Profit is not revenue. When a report says that Exxon made $60 billion in the fourth quarter of such-and-such a year, most of the time it is talking about how much it *sold in total*. This is usually referred to as "sales," "sales revenue," or simply "revenue." A company could make $100 trillion in revenue and not make a cent of profit. It could even lose money on that level of sales. Revenue, then, is not profit.

> **PROFIT VS REVENUE**
>
> Profit is not revenue. **Revenue** is the total amount a company sells over a period of time. **Profit** is what is left of revenue once all costs are taken out.

Profit is the difference between revenue and total costs. Let's take a look at Exxon's sales, costs, and profit from the fourth quarter of 2019 to illustrate how profit is determined. In that quarter, Exxon reported a little over $63 billion in sales.[25] This may seem like a lot, but keep in mind, this is a global oil and gas business.

Think about how many people around the world use oil and gas. You would expect a high level of sales for a company like this. Before we can determine profit, however, we must know its costs.

FIGURE 5.1 - BIG FIRMS, BIG COSTS

Oil rigs are an expensive cost for firms like Exxon.

Exxon, like all companies, reported several kinds of costs. First, it had the cost of the oil and gas it sold. This stuff doesn't just jump into the gas pumps on its own; it must be found in the ground, brought to the surface, processed into a useable form, and transported all over the world. Exxon's cost of goods sold (that's the accounting term) was about $50 billion in the fourth quarter of 2019. The difference between this and revenue is $13 billion. This is called "gross profit." It is not the final profit, however (see Graph 5.1 for a visual breakdown of Exxon's cost and profit percentages).

In addition to the cost of the oil and gas, Exxon also had the cost of its sales force, administrative costs (accountants and office staff), and general business expenses (rent, utilities, and supplies). This totaled around $3 billion. Add to this

GRAPH 5.1 - UNDERSTANDING BUSINESS COSTS & PROFIT

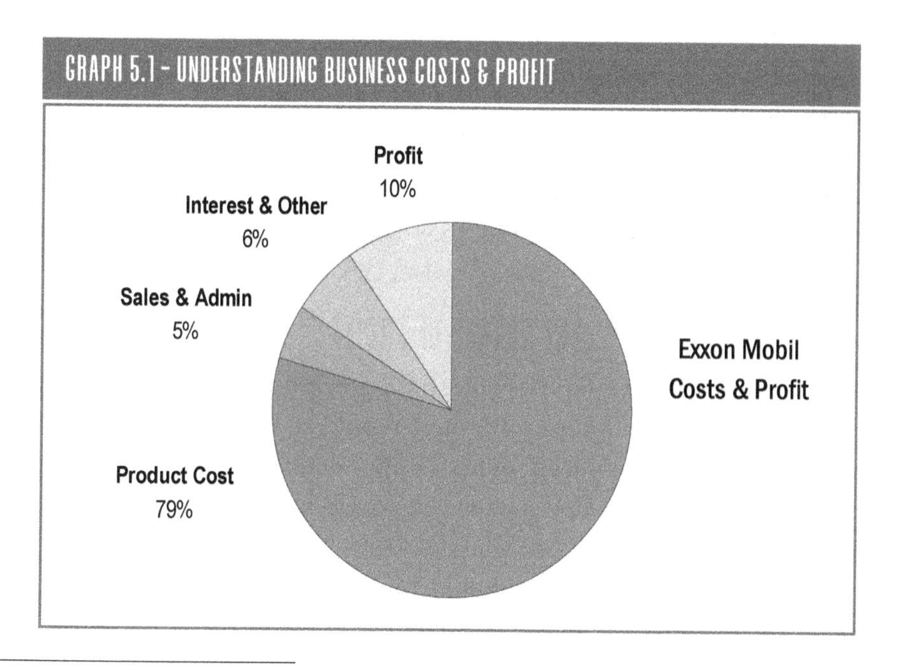

25. MarketWatch.com, "Exxon Mobil Corp."

interest on loans and other expenses, and Exxon ended the quarter with a "net profit" (as opposed to gross profit from the previous paragraph) of about $6 billion, or about 10%. So, of the $63 billion of oil and gas Exxon sold, over 90% was lost to the expenses of running the business. That profit percentage is in line with the average for America's 500 largest companies (called the S&P 500) in that same quarter.[26]

TABLE 5.1 – PROFIT MARGIN & TAX RATES 2019

Company	Profit Margin before Taxes	Tax Rate Paid
Amazon	5%	16%
Apple	24%	16%
Coca-Cola	27%	18%
Exxon Mobil	8%	26%
General Motors	4%	14%
Home Depot	14%	22%
JP Morgan Chase	31%	18%
Microsoft	34%	10%
Wal-Mart	4%	24%

IT'S TOUGH TO MAKE A PROFIT

Making a profit is a hard thing to do. A firm must convince customers of its value, keep its prices as high as it can, and work to minimize its costs. And all of this in a market that is constantly changing, with new competition entering with better products and services, a revolving set of government regulations to manage, and unanticipated environmental factors coming up at a moment's notice.

While the perception is that all businesses, especially the big ones, make tons of profit every year, as we see in the Exxon example, as a percentage, profits in most cases are lower than the owners of the firm would desire, given the amount of investment, effort, and risk they must take to generate it. (Table 5.1 gives the profit margins and tax rates for select U.S. companies in 2019.)

THE HATRED OF PROFIT

Many Americans have developed a hatred of profit, as if it in and of itself was inherently evil. But without profit, there would be no economy at all, and so much of the good we experience would simply not exist. Without profit, no one would ever start a business—that means you would have to make or grow everything you need yourself. Each person, couple, or family would have to be fully self-sustaining. And, because there would be no businesses, there would be no Coca-Cola, Apple, Ford, Boeing, or Wal-Mart.

---

26. CSI Market, "S&P 500 Profitability."

This may seem like a silly point to make, but I do it to show that profit is not something added to an economy by greedy businessmen. It is an essential and irreplaceable component. Profit, therefore, is not bad; it is mandatory.

Furthermore, think about what comes out of profitable companies. Jobs, for one. A company making a profit is paying workers (and likely hiring more) so it can keep making that profit. Exxon employs 74,900 workers[27], Wal-Mart employs 2.2 million[28], and Amazon has 798,000 on staff[29].

And then there are the benefits. A profitable company provides critical benefits for its employees, like medical, dental, tuition reimbursement, and retirement. If you are not making a profit, or if you are making a very small one, you cannot afford to dole out expensive benefit packages.

I have always told my children they need to work for companies that make a lot of profit, because they can afford to give them the perks they want. Apple, for example, in the fourth quarter of 2019, had revenue of $92 billion and a final profit of $22 billion, a whopping 24% (compared to Exxon's 10%). Apple is known for having some of the best benefits in the world, including discounts on its products, extensive family leave before and after childbirth, and free, company-sponsored concerts with artists like OneRepublic.[30] It can only afford those benefits because it makes a lot of profit.

And where do you suppose government welfare and retirement benefits come from? Leaving off for now the amount the government borrows to pay for its programs, tax revenue comes from workers with jobs (who are employed by profitable companies) and corporate profits. Companies are not taxed on revenue; they are taxed on profits (the net profit discussed in a previous section). Even social security benefits are paid for out of payroll taxes (part of which is paid by the employee and part by the employer).

TABLE 5.2 – THE BENEFITS OF PROFITS
1. **Profit provides jobs with medical and retirement benefits for employees.**
2. **Profit provides taxes to pay for benefits to the poor and needy.**
3. **Profit provides taxes used to pay social security and military retirement benefits.**

On the point of social security, in spite of popular opinion, there is no social security bank account where the payments of prior generations are

27. Forbes, "ExxonMobil (XOM)."
28. Macrotrends, "Walmart: Number of Employees 2006-2020/WMT."
29. Tugba Sabanoglu, "Number of Amazon.com employees 2007-2019." *Statista*, 30 November 2020.
30. Aine Cain, "6 Incredible Perks for Apple Employees." *Inc.*, 22 November 2017.

stored away. The fund is only a record of what is owed; the money was spent long ago (see chapter 9 for more on this dirty little secret). Payments for today's retirees come from the taxes of those who have jobs now and the contributions of the companies they work for. Because of this, there would be no money to pay social security or any other government benefit today (like military and postal service retirement) without profit.

I have a friend who draws social security and receives military retirement benefits at the same time. What is interesting is that this person is always going on about the evil of big corporations and the rich. All the while, his entire livelihood is financed by the very people he hates. If he had his way, and the profits and wealth of these "unrighteous" people and organizations were taken away, he would lose his home and starve to death.

Profit is good, and for our own good and the good of our nation, we should want every company out there to make as much as possible.

WHY WON'T BUSINESSES STAY OPEN WITHOUT PROFIT?

"Wait a minute," someone says, "you said there wouldn't be an economy without profit. But why? Would every business really shut down if it didn't make enough profit?" Yes, it would. And it is not because of greed. If you were in their place, you would do the same thing. Let me explain why.

Like all of us, business owners want to get the most out of their time and money. Unlike most of us, however, they have risked everything they have to make their business dreams come true. Think about it, would you like it if you worked an entire week and got nothing for it? How long would you continue to do this? Or what if you took your life savings and invested it in your friend's lawn mowing business and, after five years, only got back exactly what you put in? How likely would you be to invest in his next project?

FIGURE 5.2 - EVERYONE REQUIRES THIS

Business owners require a return on their investment.

In both cases, whether you realize it or not, you are expecting some kind of benefit (dare we say *profit*) for the use of your resources, time, and labor. Well, this is exactly how business owners view their investments in a company, whether those owners are the individuals who started the firm, the executives who manage it, or the investors who buy its stock and finance its operations.

Like you, owners have options. They don't have to continue to give the company their labor or money. They can pull it out and use it somewhere else, somewhere that gives them more return on their investment.

This comparison of options—that is, what could be done with the resource if it wasn't tied up in its present use—has a name in economics. It is called "opportunity cost." Opportunity cost is what you are giving up because you made one decision over another.

When you choose to go to a movie instead of a ballgame, you are giving up the benefit you would have received at the ballgame. If you get more joy, let's say, by going to the movie that night, you made a good investment; if not, you didn't. But whatever you do, you can't choose to do both. You can only do one or the other.

Business owners have opportunity costs as well. When a man chooses to open his own business, it costs him two things. First, it costs him the income he could have earned if he had taken a regular job. If the man is a graphic artist who could earn $75,000 a year working for Disney and he instead started his own company, his opportunity cost is $75,000. It's what he gave up to start the business.

Second, it costs him the return he could have gotten if he had invested the money he used for the business in something else. Let's say he used $10,000 of his savings to start his graphic arts business. If he could have earned 10% in the stock market, for example, his opportunity cost is $1,000 (10% of $10,000). So, if he doesn't make 10% or more from the $10,000 investment in his business, he would have been better off not to start the business and just buy stocks. This second type of opportunity cost applies to anyone who puts money into a company, like investors who buy a company's stock.

What does this have to do with why owners must earn a certain amount of profit to keep their doors open? Because, if the level of profit is not greater than they could have earned somewhere else, it doesn't make sense to keep using their money and time in the business. Put yourself in their situation: Would you keep your business open if you made less than you could working for someone else or doing nothing? Of course, not.

A second reason owners require a healthy profit is the amount of work they have to put in to start, run, and manage a business, or to create the extra income they need to invest. This is an often-overlooked truth about business owners—the amount of work and risk required to create a successful firm.

OPPORTUNITY COST

Opportunity cost is what a business owner gives up to start a business. It includes his salary at another job and interest from savings used in his business.

TABLE 5.3 - WHY OWNERS REQUIRE PROFIT
1. Owners want the maximum return on their time and money investment.
2. Owners have other options for their time and money that may pay more.
3. Owners don't want to work that hard and sacrifice that much for nothing.

I know several wealthy business owners. One in particular owns a tech company in Charleston, SC. He has a house in Costa Rica and one in Florida, and he lives in a mansion in the ritziest part of town. Oh, and he drives a bunch of really nice cars, too. Can't forget that.

So, why bring him up? Because he comes in at five every morning, works almost every weekend, and is down in the trenches with his techs every chance he gets. He is constantly under stress and almost never has a minute's peace.

The point? Why would you do all of this if you couldn't make a serious profit by doing so? It's just too much work to make one or two percent and barely get by.

And think about the risk he is taking. Who replaces the tools that are stolen from his trucks, or pays for the labor to overcome a mistake his estimators made, or puts up his property to secure a bank loan to open a new office? There are literally hundreds of things a day that could go wrong and not only eat away at his profit but also risk his entire fortune. Why do this if you can't be richly rewarded for it?

People talk about hard working men like this as if they were criminals. But the real crime is to let them do all of this and then have the government come in, take what they have rightly earned, and give it to someone who has lounged around all day working the system.

WHERE DOES THE BIBLE FALL ON THE ISSUE OF PROFIT?

You may be surprised to know that the Bible has nothing against profit. Consider James 4:12-16:

> Come now, you who say, "Today or tomorrow we will go to such and such a city, spend a year there, buy and sell, and *make a profit*"; whereas you do not know what will happen tomorrow. For what is your life? It is even a vapor that appears for a little time and then vanishes away. Instead you ought to say, "If the Lord wills, we shall live and do this or that." But now you boast in your arrogance. All such boasting is evil.

Apparently, some of those James was writing to were businessmen. They had products to sell, and part of their business model was to take those products to various cities and market them. Out of this effort, they expected to earn a profit. The Greek word used here for *profit* means "gain." These men expected to gain from their sales activities in other cities, and gain is exactly what profit is.

What is interesting to note is that James does not condemn them for trying to make a profit. He almost takes it for granted that they would do this. What he condemns is their arrogance in making plans without consideration for the will of God. Essentially, he's saying, "Hey, it's okay to do what you are doing, but it is wrong to arrogantly think you can do all of these things without God's blessing. Ultimately, your business plans are subject to God's will and not your own."

Why would James take for granted the pursuit of profit? Because of many of the reasons we have already mentioned. Without profit, businesses would not exist to meet the needs of people, pay salaries, provide benefits, or pay taxes. God loves people, and he wants them to have the things they need. He has also commanded them to work to provide for their own (2 Thessalonians 3:10) and to pay taxes to provide for a capable government (Romans 13:7). If you think about it on a deeper level, the inner desire to get the most out of our resources, the desire from which the drive for profit comes, was placed in man at his creation. It is a part of God's way of assuring his survival and prosperity. Pursuing profit, therefore, is not evil in God's sight.

factors that limit profit

Today, profit is equated with greed. Everybody who tries to earn a profit is "a greedy capitalist." It's as if every businessman in the country is trying to swindle every dollar he can from his customers. Is this true?

Of course, not. As I have already said, you must earn a profit or you will not stay in business. And what happens to all your employees, stockholders, and vendors when you close your doors? Profit must exist for businesses to survive and people to prosper. It is not greed that drives profit; it is survival.

Do some people seek profit out of greed? Sure, they do. As our nation becomes more and more godless, people will do more and more corrupt things to benefit themselves. But, in a free market economy like ours, there are built in checks and balances to how much profit one person or company can earn.

In a market with as many competitors as there are in the U.S., if a company is trying to make outlandish profits, most of the time its customers will find a competitor selling that product or service for less and buy it from them. Because of this, over time, competition drives down prices. This is a fundamental law of economics.

> **GREED AND ECONOMIES**
>
> Greed cannot be eliminated by an economic system. Capitalism, however, uses it for good through incentives, and minimizes it through competition.

Let's say I come up with a great new product to cure color blindness. At first, because I am a greedy dog, I charge an extremely high price for it and make a ton of profit. (And why shouldn't I, given the investment I made to develop it and the risk I took in launching it?) Over time, other businessmen start to take notice and begin to develop their own products. They may even copy mine and make it at a lower cost.

Eventually, those cheaper products begin to take sales from me. So, what do I do? I lower my price to keep my customers (and keep making those greedy dollars). As time goes by, more and more competitors enter my market, and the profits drop so low that some competitors pull out altogether. In the long run, the price stabilizes much lower than my original price and my profits are minimized.

This example demonstrates the point that, in a free market economy, greed is not eliminated; it is minimized. There is no economic system that can eliminate greed, because greed is a part of the human heart. Greed will show up no matter what you do to prevent it.

The difference between a free market economy and a socialist economy is that the power of greedy people is limited in a free market, but it is not in a socialist system. In a socialist system, the greedy people make their way to the top and use the work of the people below them to feed their greed. The common people in most socialist societies are dirt poor, but their leaders live in utter luxury (see Figure 5.3 for an example from North Korea).

FIGURE 5.3 - GREEDIEST OF ALL

Socialist leaders, like Kim Jong-un of North Korea, live in luxury while their people suffer.

Is the Bible against greed? Of course, it is. From the early pages of the Bible, God condemns greed. The tenth commandment says, "You shall not covet your neighbor's house; you shall not covet your neighbor's wife, nor his male servant, nor his female servant, nor his ox, nor his donkey, nor anything that is your neighbor's" (Exodus 20:17). What is covetousness but greed?

The New Testament also addresses the issue of greed. Jesus encouraged us not to measure our lives by what we have: "And [Jesus] said to them, 'Take heed and beware of covetousness, for one's life does not consist in the abundance of the things he possesses'" (Luke 12:15). Later, the apostle Paul warned about the lust for riches, "But those who desire to be rich fall into temptation and a snare, and into many foolish and harmful lusts which drown men in destruction and perdition" (1 Timothy 6:9).

Greed and profit do not have to be connected, however. God's power can help a man earn a profit but not fall to greed. A child of God understands that, while he has to earn a profit to provide for his family and employees, ultimately his trust is in God to meet his needs. Therefore, he does not have to put profit first all the time. His main goal is to do good so that he may show the greatness of God. Jesus Christ taught his people to do this in Matthew 6:25-26 and 31-33:

> Therefore I say to you, do not worry about your life, what you will eat or what you will drink; nor about your body, what you will put on. Is not life more than food and the body more than clothing? Look at the birds of the air, for they neither sow nor reap nor gather into barns; yet your heavenly Father feeds them. ...
>
> Therefore do not worry, saying, "What shall we eat?" or "What shall we drink?" or "What shall we wear?" ... For your heavenly Father knows that you need all these things. But seek first the kingdom of God and His righteousness, and all these things shall be added to you.

what God is against in business

Yes, God is for profit and against greed. But he is also against other practices common to the business world (and to all human endeavors, in fact). God is against fraud. Proverbs 20:23 reads, "Diverse weights are an abomination to the Lord, and dishonest scales are not good." Ancient markets sold products by how much they weighed. That weight was often calculated by using a scale and premeasured metal weights. If I wanted to buy a pound of flour, the vendor would put a pound weight on one side of the scale and fill the other side with product until it balanced. To defraud their customers, some ancient merchants trimmed the weights so they measured out amounts less

FIGURE 5.4 - WHAT GOD HATES

God requires honesty and integrity in business.

TABLE 5.4 - THINGS GOD IS AGAINST IN BUSINESS
1. God is against fraud and trickery of all kinds.
2. God is against lying and manipulating the truth.
3. God is against evil products and services.

than advertised. God is completely against this type of trickery in business.

He is also against lying. Proverbs 12:22 says, "Lying lips are an abomination to the LORD, but those who deal truthfully are His delight." I have been in the sales field for many years, and I can tell you there is no limit to the lies salesman will tell to get a sale. The Creator hates lies, even when they further a good goal (like earning an income). Businessmen are to tell the truth about their products, their inventory levels, and their delivery schedule. If that truth hurts their ability to earn an order, so be it. Remember, ultimately, we must trust God to provide for us, not our sales figures.

A final thing that God is against in business is unrighteous products and services. No matter how many employees you support, how much you give to charity, or how much you pay in taxes, if your business is unrighteous in God's sight, he stands completely against you. God does not support any business that performs abortions; sells alcohol or marijuana; provides sexual services, including pornography, prostitution, sex lines, or stripping; or provides betting or gambling services of any kind. If you find yourself in any of these kinds of businesses, you need to close your doors and do something good with your time and resources. God is against you.

BUT HE'S GOUGING ME!

Like the rich, profit is criticized by those who do not understand how markets work. Profits are necessary to keep companies in business, to create jobs for employees, and to provide tax revenue for the government, among other things. For these reasons, God has no problem with it. However, if socialists had their way and all profit was stripped from the market, the entire economy would collapse. Once again, free market capitalism, including the pursuit of profit, is the best approach for a nation.

Now, let us turn our attention to the government setting prices when people feel they are paying too much or getting paid too little for something. How should we view these kinds of price controls? Don't stop reading now. More eye-opening truth is just around the corner.

CHAPTER DISCUSSION QUESTIONS

For a better understanding of the ideas in this chapter, work through these discussion questions in a small group:

1. Explain the difference between profit and revenue. Why is this difference important when determining how much a company actually made in a year?
2. Discuss some of the types of costs a company like Exxon might have. After its costs, what percentage profit did Exxon make in the fourth quarter of 2019?
3. Explain why we wouldn't have an economy if we didn't allow companies and individuals to make a profit.
4. What are some of the benefits of profit mentioned in the chapter? Which seems most important to you? Why?
5. Summarize why businesses won't stay open without profit.
6. What does "opportunity cost" have to do with why business owners won't accept extremely low profits and keep their businesses open?
7. What does the amount of work business owners have to put in, and the risk they take, have to do with their desire to make a profit?
8. Where does James 4:12-16 fall on the idea of profit? Why would God not have a problem with profit?
9. Discuss one of the factors that limit the amount of profit a company can earn in the U.S.
10. Explain some of the things God is against in business. Is it good that he is against these things? Why or why not?

CHAPTER SIX: PRICE CONTROLS HURT EVERYBODY

Calls for government intervention in the prices of products are increasing in America. The cry for "fairness" and "equality" has somehow found its way into the world of what we pay for things. Whenever something seems to cost "too much," rumblings of price controls ripple through the nation. What most people don't know, citizens and politicians alike, is the damage these controls do not only to businesses but also to consumers. Let's investigate the matter in more detail now.

WHAT IS A PRICE CONTROL?

A price control occurs when the government dictates the price of something instead of allowing the price to develop by the free choices of buyers and sellers. By now, you should recognize what these kinds of actions represent. Because they involve government control of the market, price controls are 100% socialist.

Price controls come in two forms. There are price ceilings, where a price is not allowed to exceed a specific amount. For example, if the government set the price of a specific prescription drug, this would be a price ceiling. And then there are price floors. Here, the government sets a minimum price that can be paid for something. Minimum wage laws are a classic example of price floors. In this case, the government is setting the minimum price a company can pay for labor.

> **PRICE CONTROLS**
>
> A **price control** occurs when the government sets the price of something instead of it being set by the free choices of buyers and sellers in the market.

The goal of a price control is to make products more affordable or to provide a minimum level of income for workers. The assumption here is that there is an unfair price level in the market, either high or low. In the case of the prescription drug example, the conclusion is that prices are too high for low-income consumers, like seniors and the poor. The government intervenes in the market to require firms who produce the drug to reduce their prices, thereby correcting the alleged injustice. With wage controls, the wages are deemed too low for low-income workers and should therefore be increased. The term used in these cases is "living wage." The idea is that the lowest paying jobs on the market should still provide enough income for a person to live comfortably.

Who could possibly be against these kinds of efforts by the government? After all, don't we all want seniors and poor families to have access to the medicine they need? And who wants to see a family live in poverty when the "rich" companies who employ them make big profits?

The question really isn't who wants others to have low-cost drugs or high wages. Of course, in a perfect world, one of unlimited resources, we'd give everybody whatever they wanted for free and pay them all a millionaire's wage. But remember, economics is about determining how to spread limited resources (money, medicine, etc.) over an unlimited set of needs. While the idea in this case is to be praised, the implementation of these kinds of price controls creates more problems than it solves, and, in the process, drives up costs and reduces the availability of the things we need. Let me show you how.

AN ANALYSIS OF COMMON PRICE CONTROL EFFORTS

Consider what actually happens when the government reduces the price of a popular prescription drug. Let's say the government forces Acme Drugs (made up name) to lower the price of its revolutionary diabetes drug SugarCure (also totally made up) from $5 per pill to $1 per pill. What are the effects of this in the market? "People save a lot of money!" you say. Well, not exactly. Let's examine the full range of effects this price reduction has. Consider, first, the impact on Acme.

Remember, a company must make a profit to stay in business. As discussed in chapter 5, profit is the difference between a business' sales revenue and its costs. Acme's sales would be the number of SugarCure pills sold times the price of each pill. Its costs would include the factories, factory workers, equipment, insurance, salespeople, chemists, and research and development staff needed to make, sell, and distribute those pills. If yearly sales were $10 million and costs

FIGURE 6.1 - A SOCIALIST TARGET

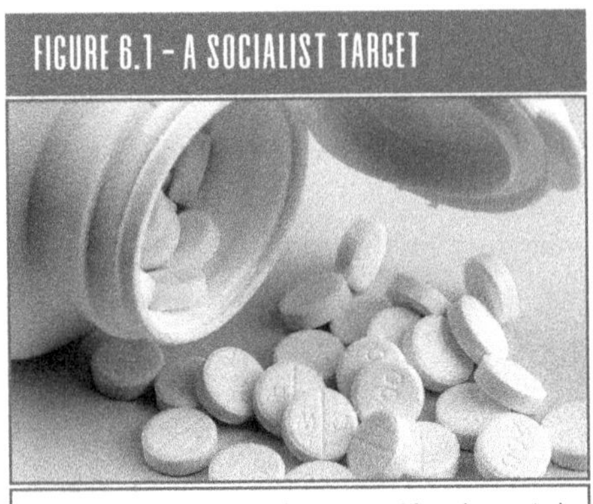

Prescriptions drugs are often targeted for price control.

were $8 million, Acme would make $2 million in profit.

But what if the price was forced to drop from $5 per pill to $1 (an 80 percent reduction)? What would total sales be at that level? Two million dollars (or 80 percent of $10 million). Costs would remain at $8 million because the company would still have to produce the same number of pills. Profit, then, would go from a positive $2 million to a negative $6 million ($2 million in sales minus $8 million in costs). What would Acme do in this case?

The first thing it would do is cut costs. If possible, it needs to get them below sales revenue so it can continue to make a profit. To do this, it would reduce spending on things like maintenance, equipment, and plant expansions. But that wouldn't be enough. Because in most businesses labor is one of the biggest costs, it would also have to lay off chemists, factory workers, and salespeople. It may even need to cut the wages of its remaining workers to make up the difference.

Let's pause here to see the real results in people's lives of a government price control. Yes, you get your cheap pills, but a whole bunch of employees lose their jobs, and a whole lot of others may take pay cuts. Ever heard about this when your favorite politician goes off on those "rich drug companies"?

No matter how much cost cutting is accomplished, however, it is hard to see how Acme could get its costs under $2 million (from $8 million before the price control). What then? Well, if you were the owner of Acme, what would you do? One of the things you would probably do is ask yourself where you could use your resources to make more money. There's no point in using your factories, workers, and expertise in a market where you can't make a profit.

Over the next few years, you would shift your focus to some other drug that doesn't have a price control strangling its profit margins. You may make a heart or cancer medication, or you may shift to making other products for the diabetes market, like foot creams or bandages. It really doesn't matter what the product is as long as you can make it efficiently and make a profit selling it.

When you pull your resources out of making SugarCure to put it into something else, what will happen to the supply of SugarCure? It will begin to dwindle, won't it? The thousands of pills you were making before would simply not be available, and because there is no profit in the market due to the price control, no other drug company is likely to jump in and fill your shoes. As a result of lower supply levels, customers would still pay $1 for the pills they purchased (because the price control is still in place); they just wouldn't be able to get as many pills as before, or they would have to wait a long time to get them (economists call this a "shortage"). Now, how do you think people are going to react to that?

TABLE 6.1 – PRICE CONTROLS' EFFECT ON AN INDUSTRY
1. Lost jobs: Price controls can cause massive cost cutting, including employee layoffs.
2. Lost suppliers: Price controls cause shortages by pushing producers out of the industry.
3. Lost long-term supply: Price controls cause other manufacturers to stay out of the industry.
4. Lower quantities: Price controls reduce the supply of the product, increasing wait times.

And consider how this affects the thinking of other drug companies. How will their long-term plans for drug development be affected by SugarCure's situation? How willing will they be to invest the millions it takes to develop a new drug if they can't recoup their investment and make a good profit? It doesn't take a genius to figure out that they will be much more careful about what they spend their money on in the future. When drug manufacturers lose the incentive to develop new drugs, how does that impact the health of society?

wage controls

What about minimum wage laws? What unintended consequences occur when they are raised? Suppose Benny's Burger Shack has 10 employees. Next month, the minimum wage will go from $8 to $12 per hour. What are the effects of this? "The workers get a lot more money!" someone says. Oh, yeah? Let's see.

As with Acme, the government can control the price, but it cannot control what Benny's (and every other company affected by the wage increase) does in response. Benny is in an extremely competitive market. He must battle not only the big national chains like McDonald's and Wendy's, he must also fight with a host of local and regional restaurants as well. Benny must provide great food, great service, and great prices just to have a chance to eke out a small profit. In his market, a $4 increase in labor costs (or a 50 percent increase) translates into a 25 percent increase in his total costs (labor is about 50 percent of the costs in a

FIGURE 6.2 – ANOTHER FORM OF CONTROL

Minimum wage laws are a form of price control.

TABLE 6.2 - OPTIONS WHEN LABOR COSTS RISE
1. Take costs out of profits if possible.
2. Raise prices and risk losing customers.
3. Cut costs, including eliminating employees.

restaurant). How does he make up for this additional cost?

He really only has three choices. He can take it out of his profits, but they are already so low he has very little room to do that; his profits simply aren't high enough to absorb all that additional cost. Another option is to raise his prices. Given the fact that all the other fast food restaurants have the same problem—what to do about the additional labor cost caused by the wage law—he may be able to get by with this. In other words, if, to cover the additional cost of wages, all restaurants in his market raised prices, Benny could raise his and not lose a great deal of business because of it.

Now, let's pause here and consider what just happened. The government, in order to help low-income workers, just raised the minimum wage. But, by doing so, it also raised the prices of everything made or sold by workers at this wage level.[31] So, the worker gets a raise, but, at the same time, his cost of living goes up, as it does for everybody else, making his raise not as much of a raise.

Now, imagine that Acme uses minimum wage workers in its SugarCure factories (the situation we discussed earlier). Think about what government intervention has done to Acme's business. On the one hand, the government is raising its costs by increasing the minimum wage, and, on the other, it is reducing its revenue by fixing its prices lower than the market. How long would you stay in business if the government did this to you? At some point, you would just say, "This just isn't worth it. I'm going to close the doors and find something else to do."

This is the incentive problem we talked about earlier. When socialist policies like wage and price controls are enacted, they reduce the incentive to create, build, and take risks. When this happens, everybody loses, and poverty is the sure outcome.

Benny's final choice is to cut staff. If he can't pass all the cost increases along as price increases and if he can't eat it as lost profit, he will have no choice but to fire people. What many companies are doing today to fight the high costs of labor

31. They also raise most everybody else's pay level because the minimum wage is used as a benchmark for higher hourly pay rates. Imagine if you made $15 an hour and the minimum wage was $8, and then the minimum wage jumped to $12. That $15 an hour rate doesn't look so good anymore, does it? How long before you go into your boss' office and ask for a raise? Because of this phenomenon, minimum wage increases tend to raise most other wage levels too.

is to replace workers with machines and technology. Even convenience stores (which use a lot of low-cost labor) are doing this. In my area, Parker's Kitchen, a convenience store chain, is building a lot of new stores. But they have very few staff. Self-pay kiosks have replaced workers (see Figure 6.3). Why? Because minimum wage laws have made them cheaper than employees.

FIGURE 6.3 - TECHNOLOGY VS WORKERS

High minimum wages cause technology to replace workers, as in Parker's convenience stores.

So, here we go again—government intervention producing lost jobs. Now, if you are one of the lucky few to keep your job, the pay hike is great. But if you are laid off, it's the worst news of all.

so, what's the answer?

The answer is to let the market set the prices for both products and labor. The free decisions of buyers and sellers will create the most efficient price in each case, will ensure the products we need are produced in the quantities we require, will maintain the incentive for creators of those products, and will maximize the number of jobs available. Once again, free market activity, not government intervention, provides the best outcome for everyone.

We must also see the foolishness of the idea of a "living wage." What does it take on an hourly basis to live in our modern world? There is no firm answer that applies to all people at all places in all situations. But whatever it is, it is way higher than even the new minimum wages being discussed today.

My oldest daughter just finished college with a two-year degree as a paralegal. She was blessed by God to get an entry level job as a legal assistant at the biggest law firm in Charleston, SC, where we live. Her starting wage was $17 per hour, and she still cannot afford to live on her own (and she owns her own car and has no debt, student or otherwise). I would estimate that she would need $22-25 per hour to get out on her own (and this is in an area with a relatively low cost of living). What

A "living wage" would cause prices on everything to rise drastically, harming the economy and costing jobs.

FIGURE 6.5 - THE LOW WAGE ANSWER

The answer to low wages is personal responsibility, not government intervention.

happens to prices if this becomes the minimum wage? How many jobs will be lost at that number?

The solution, as I have said before, begins with lower income workers taking responsibility for their own lives. They must take advantage of the educational and training programs available to increase their skills and value in the market. The answer is personal responsibility, not government intervention.

For the truly poor (those who cannot pull themselves up through their own effort), as long as funds are available, programs addressing their needs must be developed and effectively managed. But, as I will show later, the U.S. government is out of money, and, therefore, expensive programs for the poor will be forced to diminish in days to come. Perhaps then the churches will once again carry out their God-given responsibility to help the needy.

WHY DO PRICES GO UP?

There is a great deal of confusion today about why prices change. Every time prices go up, people think the rich corporations are taking advantage of them. But this isn't the case. There are several reasons prices go up (and down) all the time, and none of them are criminal. The first is a change in demand.

demand and price changes

Imagine you start building wooden picnic tables to sell online. Your first three tables are ready in April, so you put them online for $150 each. To your delight, they sell in a few hours. So you make three more. This time you raise the price to $175, and they sell just as fast. You do this for a few more rounds, and eventually the price tops out at $225 each. By then, you've sold as many as you want to build, so you take the summer off and plan to start again in the fall.

When the fall comes, you put your next three online at $225 and get no bites. Not wanting to lose your investment, you drop your price to $200 and then $175. Eventually, you sell them for $150. You make a few more, but, as the winter sets in, you can only get $125 for them.

So, what happened here? Why did you get different prices at different times of the year? And why could you raise your price at one point in time but had to lower it at another? The reason prices changed is because demand changed. In the spring, people are gearing up for the summer. Picnic tables are in high demand because people want new ones for backyard cookouts and parties. In the fall, however, people still want them but not as badly, so they are not willing to pay as much. As fall winds down, there aren't many people thinking about picnic tables anymore, so the prices fall even further.

What's the lesson here? The lesson is, when demand shifts, prices shift. Higher demand times mean prices go up (to a certain point), and lower demand times mean prices go down (also to a point). It's basic economics.

Now, let me ask you a question: When you raised your price in the spring, were you taking advantage of people? Were you being greedy and trying to extort as much money from picnic table buyers as you could? No, you weren't. What you were trying to do is maximize the return you received from the time, money, and risk you took in making those tables. It is only natural—and wise, I would add—to get as much as you can from your efforts. Only a fool would do all that work and then settle for as little return as possible. This idea, that we all try to make the most from our investment, is called "profit maximization."

Let's apply this to a big company like Exxon. I've often heard people complain about the increase in gas prices around the July 4th weekend. In their minds, the greedy gas companies are just taking advantage of them by raising prices when they are taking family trips. So, how does the previous example help explain those price increases?

During high travel holidays, like July 4th, demand for gas goes way up. A lot of people are hitting the road to go to the beach or the lake or the golf course. During these times, people are not as sensitive to the price of gas as they are at other times (this is called "price elasticity of demand"—yeah, I know, a real mouthful). In other words, on a normal day, if they saw the gas price was higher at a certain gas station, they might just pass on by to see if the price was lower somewhere else. Or they may just wait a few days to see if the price would come down (if their tank was not too low). On holiday travel weekends, however, they don't do this. They pull in and fill up, pretty much regardless of the price.

FIGURE 6.6 – WHY PRICES GO UP, PART 1

Gas prices go up on holidays because of high demand.

Exxon (and all the other gas companies)

DEMAND & PRICE CHANGES

When people have a greater desire for a product (demand), they are willing to pay more. Over time, suppliers raise prices until demand and supply balance.

know this. To maximize their return, then, they raise the prices slightly to capture additional profit. "Wait a minute!" you say. "That's unfair!" No, it isn't. Just as you were not wrong to raise the price of your picnic tables to maximize your return, Exxon is not wrong to raise its prices to do the same. They are only doing what each of us would do in the same situation. Furthermore, don't forget that you don't have to buy from them. No one is forcing you to take that trip or to buy gasoline from Exxon. If you choose to do so, you can hardly blame them. Personal responsibility, remember?

This brings up another important point. Because we live in a free market, where people are free to start other gas companies, and you and I are free to buy from whomever we want, Exxon can only raise its prices so much. If it goes too far, it opens the door for its competitor on the next corner to beat its price and take its business. So, the market itself limits how much of an increase there will be at high demand times. This is why you see Exxon's profit margin at 10 percent. The market tends to limit the profits companies can make.

Sometimes demand works in the favor of companies like Exxon, and sometimes it does not. There are not only high demand times; there are also low demand times. During off-peak seasons, the dead of winter, for example, gas prices tend to fall. The gas companies know you don't need gas that badly, so they lower their prices to try to get you to buy more. On these occasions, their profits are much lower, sometimes even under their costs. It is interesting to note that the people who complain about gas companies in the summer don't have a word to say when those same companies lose their shorts in the winter. Funny how that works, isn't it?

Before I close this section, let me also mention that the supply of goods affects price levels too. People don't complain about this as much, so I won't spend as much time on it. In the case of gas prices, whenever the supply of gas is reduced (or increased) significantly, the price will change. Hurricanes in the Gulf of Mexico often cause gas prices in the southeast to go up. Storms can damage oil rigs, reduce the flow of ships, and curtail production. When this happens, prices usually go up until the effects of the storm have been corrected (this is known as a "supply shock").

limited resources and price changes

Demand has a big impact on price fluctuations. But there is one other key reason prices change, one that is good for all of us in the end (not just for companies trying to maximize profits). In chapter 1, I gave the definition of an economy as the way a nation satisfies the endless needs of its people with the limited resources it has. The price of products is the primary way these limited resources are distributed fairly among those endless needs. Let me explain how this works.

FIGURE 6.7 - WHY PRICES GO UP, PART 2

Gas prices go up after storms because of low supply.

We can all agree that, as big as the gas companies are, there is only a limited amount of gasoline to go around on the holiday weekend. Make no mistake, by the grace of God, there is a lot, but it is not an infinite amount. I think we can also all agree that, at any one time, some people are going to need gas more than others. Some are taking trips over the holiday, trips they've planned all year, and some are just chilling at home and couldn't care less. Finally, we should all be able to agree that people tend to buy more of something the less it costs, and less of something the more it costs. With these principles as a backdrop, let's look at why it's best that gas prices go up during July 4th weekend.

Imagine gas prices before the holiday were $2.00 a gallon. What would happen if Exxon lowered its price to $1.50 a gallon for the holiday? Would people buy more or less gas? More, of course, because it's cheaper. What if Exxon lowered it to $0.50 on July 2nd? Everybody and their brother would fill up every can they could find to take advantage of the epically low price. At that price, even the guy sitting on his couch for the holiday would get up, drive a mile to the store, and fill up.

Now, let's say your family gets a late start on your trip to Myrtle Beach on July 3rd. Before you can get the kids ready, pack all the suitcases, and make it to the local Exxon, it's two o'clock in the afternoon. Even though you had hoped to leave earlier, your whole family is still super excited to take the trip you planned and sacrificed for all year. You pull up to the pump, jump out to fill up, and what do you see? A sign plastered on the pump saying, "Out of gas." Now, what are you going to do?

This slightly unrealistic example helps demonstrate an important point. Price tends to limit how much people buy of something, and, therefore, ensure only those who need it most get it. Because of the trip you planned, you were

THE RATIONING EFFECT OF PRICE

Higher prices ensure that the people who need a product the most get it. They also prevent wasteful buying by those who do not need it at the time.

willing to pay more for gas than someone else. But, because the price was reduced too much, people who did not want it as much bought it anyway, and you were left without. If Exxon had raised the price to, say, $2.50, it would have discouraged those who didn't really need gas at that time from buying it, and there would have been gas for you when you got to the station. This phenomenon is called the "rationing effect of price."

We can learn another important lesson about economics here. And that is that fairness in an economy is not defined as a situation in which everybody gets the same thing everybody else has, in the same quantity. Fairness occurs when products are distributed to those who need them the most and are willing to pay to get them. When this happens, an economy is called "efficient," because its limited resources are given to those who need them most based on their willingness to pay for them.

Does that mean the poor get nothing? Of course not. As shown in chapter 4, in America, the poor live better than anywhere else in the world. A functioning, compassionate, and abundant society should provide for the legitimate poor. But it will only have the resources to do so if it follows wise economic principles, like raising prices when demand is high.

BUT I WANT IT FOR FREE!

It sounds good in theory, but when the government tries to control prices, as socialism advocates, it ends up hurting everyone involved, consumers, businesses, and workers alike. This is a common outcome of all socialist policies. The best solution, therefore, is to let the free actions of buyers and sellers decide the price. This ensures enough profit for companies to keep producing and enough supply so that people who need those products can get them.

But low prices are not the goal of an increasing number of Americans. They want all their stuff for free, and they want the government to pay for it. How does this desire stack up to true economic wisdom? Chapter 7 will reveal all.

CHAPTER DISCUSSION QUESTIONS

For a better understanding of the ideas in this chapter, work through these discussion questions in a small group:

1. What is a "price control," and what is its goal?
2. The chapter discussed the hypothetical situation where Acme Drugs was forced to drop the price of its Sugar-Cure diabetes medicine from $5 per pill to $1 (an 80% reduction). Based on this discussion, answer the following questions:
    a. Would any of Acme's cost-cutting measures (to get it back to making a profit) affect its employees? Why would this be the case?
    b. Over the long run, would Acme shift from making Sugar-Cure to some other product that did not have a government price control? Why or why not?
    c. How willing will other drug companies be in the future to develop drugs that could be subject to a government price control? Why?
3. Why can't the owner of Benny's Burger Shack just raise his prices to cover the increase in the minimum wage from $8 to $12 per hour? (Hint: Think about how Benny's customers would react to this price increase.)
4. What is the effect on customers when businesses raise prices because of an increase in the minimum wage? (Hint: Do prices go up or down for them?) What does this say about who really pays for an increase in the minimum wage?
5. How does Parker's Kitchen use technology to respond to continued increases in labor costs? What does this do to the jobs available for people?
6. Why is the "living wage" not a good idea for the economy?
7. Why did the prices of the table maker in the chapter go up and down with the seasons? Was this fair or unfair? Why?
8. Explain why prices go up during the July 4th weekend? Is this fair or unfair? Why?
9. Why do people not complain during off-peak times when gas prices are lower? What does this say about our understanding of how demand affects prices?
10. Why is it best that gas prices go up on July 4th weekend, in terms of making gas available for those who need it most? Explain why this is so.

CHAPTER SEVEN: NOTHING IS FREE. EVER.

A recent internet search revealed a shopping list of free stuff from the government that would make a Hollywood actress blush. If you're willing to invest the time, you can get free money for education, free money for a down payment on a house, free money for childcare expenses, free money for healthcare insurance, free money for electric bills, and free money to winterize your home. You can even get free cell phones, free food, and free home improvements.[32] The government has so much to give away, it even advertises (see Figure 7.1)! There's hardly a product or service available on the market today that the government isn't giving away for free.

FIGURE 7.1 - FREE STUFF EVERYWHERE

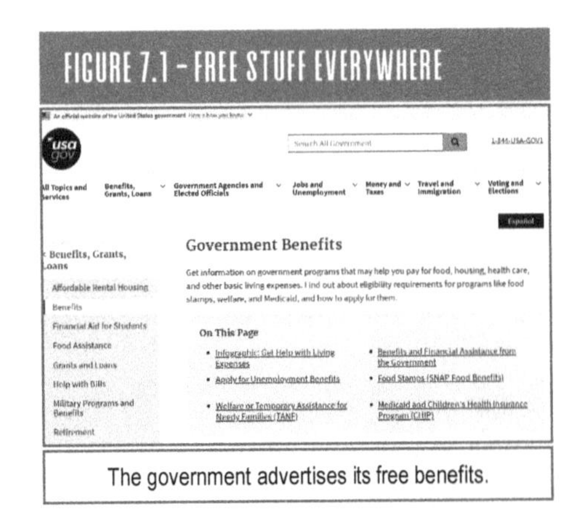

The government advertises its free benefits.

Sound too good to be true? Well, it is. In reality, all this free stuff must be paid for by someone, and that someone is a lot closer to home than most people realize. Moreover, giving everything away causes a number of other long-term problems.

Before I dig into these issues, let me address the connection between free stuff and socialism. One of the goals of socialism is to cause

32. Stephanie Colestock, "18 Ways to Get Free Money from the Government." *DoughRoller*, 24 November 2020.

people to depend on the government and not themselves. Every program the government offers for free, then, is another way for it to create that dependence and increase its power.

IS IT REALLY FREE?

Free government stuff is not free in an absolute sense. Someone must pay for it somewhere along the line. Let's use "free" money for education, specifically, subsidized student loans, as a way to understand this.

The U.S. Department of Education offers interest free loans to low-income undergraduate students. If you are getting a four-year degree or attending a trade school at least half time, you can borrow money for your education without paying interest. From your perspective, then, this money is "free."

But is it really free? It is only free for you; for the government, it is not free at all. Every month, when your interest payments come due, the government writes a check to pay them. This is an important point, and it applies to everything we get for "free," whether from the government or anyone else. Free loans for college, free cell phones, free samples at the grocery store, you name it—if you get it for free, someone else is paying the bill.

"But it's just government money," someone says. "Who cares?" Many people think the government has an unlimited supply of money. But just where does it get that money? There are two major sources of income for the government, taxes and borrowing.[33] All spending must be paid for out of the taxes the government collects from its citizens or the loans it takes out from its creditors.

In the case of taxes, your free loan costs some other citizen a part of his income. He works hard every day to make that money, and then the government takes it to give it to you (in the form of interest payments on your loans). How do you think he feels about that? Now, you might not care about this either because, as we learned earlier, most taxes are paid by the top 50% of wage earners. But, no matter how you feel about it, your "free" loan is not free; someone else is covering the cost.

What about when the government borrows the money? Who has to pay for this? I'll discuss govern-

GOVERNMENT'S ONLY TRUE INCOME

Because borrowing must be paid back with taxes, the government's only real source of income is taxes. Borrowing, then, leads to higher taxes.

33. The government also receives money from things like tariffs, customs duties, and fees, among other things, but these are small in the grand scheme of things. When the government spends money, it gets the bulk of it from tax collections or borrowing.

ment debt in more detail in chapter 9. Right now, let me just give you the big picture. When the government needs to borrow money, it doesn't go to a bank to take out a loan like the rest of us. It sells bonds to raise the money it needs.[34]

How does this work? The government sells a bond for a certain amount (say $100). It then uses that money to do whatever it wants to do. Down the road, when the bond "matures" (or comes due), it pays the bond holder a little more than he paid for the bond to start with (say $102). This additional amount ($2 in this case) is called the "interest" or "yield."

When the government borrows money to pay your student loan interest, it must not only pay back what it borrowed; it must also pay back the interest on what it borrowed. In case you're interested (I know, another bad pun), in 2019, the U.S. government borrowed about 23 cents of every dollar it spent (see Graph 7.1).[35] This means that, if it paid any student loan interest for you in 2019, it borrowed 23% of those payments. Ironic, isn't it? The government pays interest on the money it borrows to pay your interest. Only in America.

And where does the government get the money to pay this interest? Eventually, it must come from the money it collects in taxes. You can't borrow forever; at

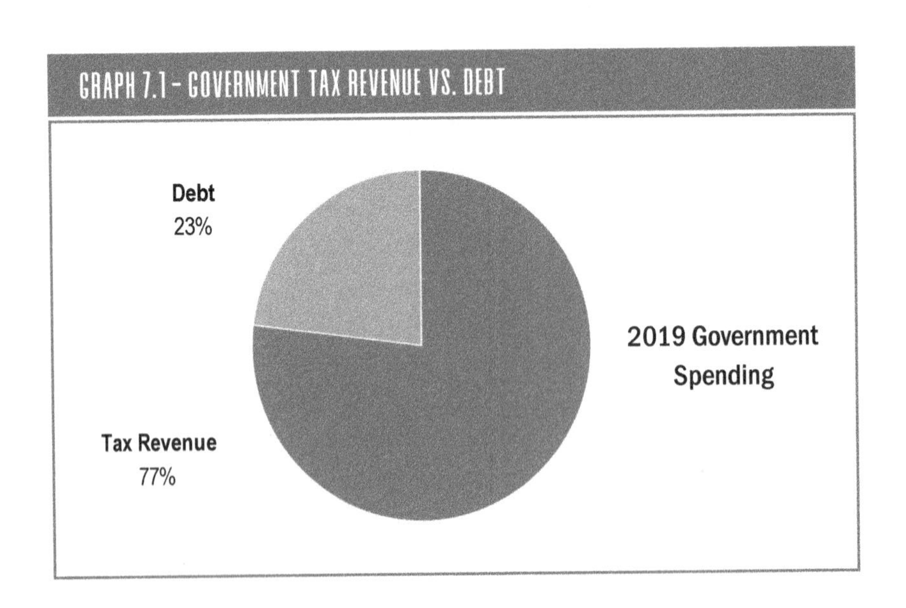

34. Technically, the government sells three types of debt instruments: treasury bills, treasury notes, and treasury bonds. Treasury bills come due in a year or less, notes between one and 10 years, and bonds from 10 to 30 years. Here, we'll just use the generic term *bond* to refer to all three types.
35. Congressional Budget Office, "The Federal Budget in 2019: An Infographic." 15 April 2020.

some point, you must pay back your loans with your own money. The only money the government gets that it can call its own is taxes. So, whether the government pays your interest payments with tax money or through loans it gets from others, ultimately the taxpayers must pay for your "free" loan.

This truth—that all free stuff from the government must eventually be paid for from tax revenue—should make all of us concerned about how much our government gives away. You may think that because you are not "wealthy" (in the top 50% of income earners, in other words) that this discussion really doesn't affect you. But here's the kicker: As the government's debts pile up, at some point everybody's taxes, from the top 50% to the bottom 50%, are going to rise drastically to repay those debts (see chapter 9 for more on this). This means that all of this free stuff will eventually cost *you* a whole lot of money.

Starting to care now?

FREE FROM ANOTHER ANGLE

Let's look at this from another angle. Let's say your student loans are not interest free. You must pay the interest on them yourself. And let's say that something happened that caused the economy to slow down drastically (say, a pandemic) and, instead of the government paying your interest, it told the banks holding your student loans they couldn't charge interest for the next six months. Because of this, in effect, you have a "free" loan (at least for the next six months). Is this money really free, or does someone else have to pay for it?

To answer this, you must first remember that banks are not charities; they are for-profit businesses. The owners of banks will only continue loaning money as long as they can make a reasonable profit on those loans (see chapter 5 for why). The primary source of profit for banks is the interest they charge on the money they loan out. So, how would they respond if the government told them they could not collect interest for six months?

In the short-term, a couple of things would happen. If the bank held only student loans, or if a large enough portion of its lending was to students, it may go out of business. The bank uses the profits from loans not only to pay its owners, but also to pay its employees, rent, and utilities. Without profit, it would not be able to pay these and would be forced to close. In that case, everyone who worked at the bank

FIGURE 7.2 - NOT FREE HERE

Banks cannot give free loans and stay in business.

would be out of a job. In addition, any person or company providing products or services to the bank would lose that income. So, the minority-owned cleaning service would lose a customer; the local heating and air firm would see profits fall; and the family-owned office supply store would see revenue shrink. These firms, in turn, may lay off people to make up for that lost business.

If the bank had enough business in other types of loans (say mortgages, car loans, or credit cards), it would have to raise rates on loans in those areas. This would help offset the losses from the student loan portion of its business. As a result, mortgages, car loans, and credit cards, among other things, would cost more. Anyone borrowing from that bank would pay more for the money they need, and all of this because the government is trying to help you.

In the long-term, some banks would probably stop offering student loans altogether. They would not have a choice. To protect themselves from future losses, they would move their business to areas where the government did not try to intervene.

Finally, because of the government's intervention, banks not previously offering student loans would be hesitant to enter the student loan market. And the ones who stayed in the student loan business would likely make their loans harder to get and raise the rates on the loans they did make. This would mean fewer student loans for those who needed them, and those who got them would pay more for them. What effect would this have on education in America?

DOES THIS SOUND FAMILIAR?

If you are paying attention, you have probably noticed that this kind of government intervention—that is, telling banks they can't charge interest—is just another example of price controls we discussed in chapter 6. The reason I bring it up here is because we are seeing more and more of this by the government, especially in response to bad economic times (like suspending rent payments, forbidding evictions for not paying your mortgage, and other "protections" for consumers).

The goal is not a bad one. The government is, in essence, trying to protect people from the suffering that comes when the economy makes a turn for the worse. But, as you can see from this discussion, there is always suffering when the economy slows down. In the case of student loan interest, helping the student by suspending interest payments on his loans hurts the banks, their employees, their suppliers, and consumers at large. In an economic downturn, it is not a choice of suffering or not suffering; it is a choice of who suffers the most. So, who should suffer when the economy falls?

A fair rule (and one that produces the least long-term economic suffering) is that each person should suffer based on his own choices and his own area of responsibility. If you chose to take out a student loan, you are responsible to make the payments even when the economy goes south. You cannot go to the government for relief when you can't make those payments. Similarly, if a bank chose to lend you money and you cannot repay it, the bank is responsible to deal with the lost income that comes from your non-payment. It cannot go to the government for relief based on a decision it made. If we do not allow these consequences to fall as they naturally would, we will make the problem worse down the line and bankrupt our government in the process.

WHO SHOULD SUFFER?

When the economy falls, each person or organization should suffer based on its own choices and area of responsibility. The government should not bail anyone out.

OTHER PROBLEMS WITH MAKING EVERYTHING FREE

The issues mentioned above are not the only problems with the government providing more and more products and services for free. Three additional problems are worth noting: the problem of waste, the problem of quality, and the problem of overuse. To illustrate these, consider the example of the Affordable Care Act, also known as Obamacare.

As with the interest free student loans, Obamacare is not free. When the law was passed in 2010, the Congressional Budget Office (the part of the government that estimates the costs of things the government does) estimated its cost at between $940 billion and $1.8 trillion from 2012-2022 (or $94 to $180 billion per year).[36] Supporters claimed these costs would come down over time, and eventually Obamacare would pay for itself. But according to Charles Blahous, a researcher at the Mercatus Center at George Mason University, these reductions never took hold.[37] In fact, in 2017, Blahous stated that the federal government could save anywhere from $228 billion to $1.07 trillion over an 8-year period if Obamacare was repealed (between roughly $29 and $134 billion per year). "Repealing

FIGURE 7.3 - NO "FREE" HEALTHCARE

Obamacare costs the government billions every year.

36. Debt.org, "ObamaCare and Associated Costs."
37. Charles Blahous, "The Fiscal Effects of Repealing the Affordable Care Act." *Mercatus Research*, 4 April 2017.

the Affordable Care Act," he said, "considered separately from its possible replacement, would substantially reduce future federal budget deficits."

Strictly speaking, Obamacare is not free healthcare for everyone; it is free for low-income consumers. What Obamacare tries to do is force everyone to buy health insurance so that those who can afford it pay for those who cannot. As the numbers above indicate, the system is not working, because the government must pitch in between $29 and $134 billion a year to cover its costs.

But let's say that free healthcare for everyone was our goal. What problems other than cost would plague such a system? First, consider the problems of waste and quality. Government programs have a difficult time keeping costs down and revenues up. In economic terms, they are *inefficient*. Consider the U.S. Postal Service as an example. According to the Government Accountability Office, the Postal Service lost $69 billion from 2007 to 2018.[38] Moreover, due to the huge level of debt of the U.S. government, it could be argued that *every government program* is inefficient (because every year the government pays more for them in total than it takes in). Given this reality, if the government ran the healthcare system in the U.S., is there any reason to believe it would perform any better? No. Like everything else the government touches, over time, its costs would spiral out of control and taxpayers would be left to make up the difference.

Quality also suffers when the government takes over an industry. What kind of service do you get at your local DMV, social security office, or post office? No, it is not always the worst it could be, but long lines, foolish rules, and confusing processes are not uncommon. What about government-run Veterans Administration (VA) healthcare services? According to a 2019 USA Today report, while a majority of VA hospitals reported lower death rates than private facilities, many had higher rates of preventable infections and bed sores (both signs of neglect), and "nearly every VA performed worse than other medical providers on industry-standard patient satisfaction surveys."[39]

FIGURE 7.4 - A CONTINUAL BIG LOSER

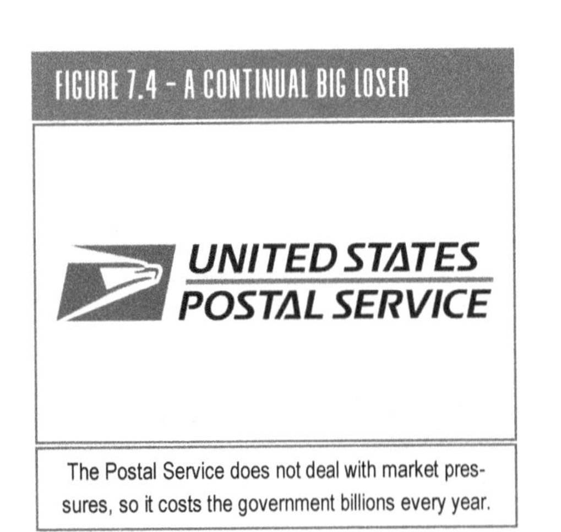

The Postal Service does not deal with market pressures, so it costs the government billions every year.

Both of these problems stem from the fact

38. Government Accounting Office, "U.S. Postal Service's Financial Viability - High Risk Issue."
39. John Kelly, Jim Sergent, and Donovan Slack, "Death rates, bedsores, ER wait times: Where every VA hospital lags or leads other medical care." *USA Today*, 7 February 2019.

that government programs do not have to face the full effects of competitive market forces. The Post Office, while in competition with the likes of UPS and FedEx for package delivery, has a monopoly on letter mail delivery and can always go back to the government for more money if expenses exceed revenues. The VA, on the other hand, has a complete monopoly on veterans. If a veteran doesn't like the service he gets from the VA, his only option is to go to a commercial healthcare provider and pay for those services himself. Not a great option for most.

One benefit of the market is that it tends to reduce waste and improve quality because it places firms in a fight for survival. Companies in a competitive market cannot allow their costs to exceed their revenues for long. If they do, they will go out of business. Therefore, when a competitive firm has too many employees, it lays some off; when it has too many buildings, it gets rid of some; and when a product is not selling, it cuts it from its offering. Similarly, competitive firms cannot allow quality to fall in the eyes of their customers. If they do, the customer will quickly take his business elsewhere and the firm's revenue will fall. If this continues for too long, again, it will go out of business. Unlike government programs and departments, there is no rich benefactor standing in the shadows waiting to bail them out (namely, the taxpayer).

A third problem with the government providing products and services for free is the problem of overuse. When you give someone something for free, they will use more of it, even if they really don't need it (remember the rationing effect of price?). Charging someone for something is not only a way to recoup its cost and provide a profit; it is also a way to make sure it goes to the ones who need it most.

Regardless of what resource you are talking about, there is only so much to go around. We must, therefore, use what we have in the most efficient manner possible. This wise rule goes out the door when the government provides it for free.

I have a pastor friend who lived in Canada for several years. He once told me something about Canadian healthcare that illustrates what we are talking about here. He said, "Yeah, the healthcare in Canada isn't bad, if you can ever get in to see somebody." His point was that, yes, Canada has "free" healthcare, but because it is free, every-

TABLE 7.1 - THREE PROBLEMS WITH FREE STUFF
1. **The problem of waste: Government programs are inefficient and tend to waste resources.**
2. **The problem of quality: The things government provides are of lower quality than the market.**
3. **The problem of overuse: When people don't have to pay for things, they use them too much.**

FIGURE 7.5 - "FREE" BUT YOU MUST WAIT

Patients of "free" Canadian healthcare wait months to see doctors and have procedures performed.

body wants it all the time, making wait times skyrocket.[40]

Research supports this. The Fraser Institute reported that the average wait time in Canada to receive treatment from a specialist was nearly 21 weeks in 2019 (more than five months).[41] The wait time to have a consultation with a specialist (not actually receive treatment) was just over 10 weeks (two and a half months). In the U.S., 70% of patients see a specialist in less than four weeks, and 61% had a procedure completed within a month.[42] One of the obvious reasons Canadians must wait so long is that healthcare demand is high because nobody has to pay for it.

HOW TO FIX THE "FREE" STUFF PROBLEM

As the previous examples illustrate, the move in America for the government to provide more and more free stuff is not sustainable. The Federal government is running constant deficits, and borrowing to pay for free products and services increases interest costs and places higher burdens on future taxpayers. But what if someone really needed those products and services?

If you need the things the government is now giving you for free, you must begin to provide them for yourself. You must take responsibility for your own needs and stop looking to the government to provide everything for you. This means you must invest more in yourself so you can earn a higher income, work more hours to generate the money you require, or learn to say no to yourself.

The Bible commands us to earn our own way. The apostle Paul said this about who is responsible to provide for our needs: "But if anyone does not provide for *his own*, and especially for those of *his household*, he has denied the faith and is worse than an unbeliever" (1 Timothy 5:8). A Christian is to work to pro-

---

40. Wait times are also high because doctors' profits are limited, creating a shortage of doctors. This is an example of rational self-interest we discussed in chapter 1. When people can't make much as doctors, they do something else instead, or take their services somewhere that pays better, like the U.S.
41. Bacchus Barua and Mackenzie Moir, "Waiting Your Turn: Wait Times for Health Care in Canada, 2019 Report." *Fraser Institute*. 10 December 2019.
42. Kevin Pham, "America Outperforms Canada in Surgery Wait Times—And It's Not Even Close." *Foundation for Economic Education*, 17 July 2019.

vide for the needs of his family. He is not to look to the government, or anyone else for that matter, to pay his way. If he does not, according to Paul, he stoops lower than even those who do not believe. In another place, he said it this way, "If anyone will not work, neither shall he eat" (2 Thessalonians 3:10).

Now, does this mean we will never need the help of others, or even the government? Does this mean a Christian can't accept unemployment or food stamps in a time of crisis? No, of course not. What it does mean is that a Christian's first thought is to work to provide for himself and his own, not to sit around and wait for the government to give him something for nothing.

The Bible also encourages us to say no to ourselves. Proverbs 25:16 says, "Have you found honey? Eat only as much as you need, lest you be filled with it and vomit," and Proverbs 25:27 reads, "It is not good to eat much honey." The point here is that it is not good to have too much of something. You have to learn to say no to yourself.

We must do this so we can save for the future. Proverbs 21:20 points this out: "There is desirable treasure and oil in the dwelling of the wise, but a foolish man squanders it." What's the writer saying here? He's saying you must wisely use what God gives you. If you consume everything you have, you only have yourself to blame when you have to go without later on.

HELPING THE POOR WITHOUT GOING BROKE

But how do we help those who can't earn enough to pay for what they need? Whatever the answer is, it is not to give them everything. As stated in chapter 4, this encourages people to be lazy and only amplifies the problem. Moreover, because of the overuse issue, the government's resources will not be used efficiently when they are given away; people will use up whatever is available without concern for what others need.

For those who are genuinely poor and cannot provide fully for themselves, a better way is to pay for part of what they need and make them work for the rest. For example, if a poor family needs a place to live, the government could pay a portion of their rent but not all of it. This will cause them to use the resource more efficiently and to take better care of it because they will have some of their own money at risk. This approach could apply to healthcare, as well. The government might offer to pay a portion of the poor person's healthcare insurance but not give him completely free insurance. Only in the most extreme cases should the government provide something for free (in other words, pay the entire cost).

BUT WE NEED MORE STIMULUS!

Nothing is free, no matter where it comes from. Someone else has to pay the bill. Even when times are tough, the government will do more harm than good by providing everyone with "free" stuff or requiring that businesses do the same. We also put our freedom and prosperity at risk by asking for and using these types of socialist enticements. Furthermore, free government-run programs, like healthcare, produce a lot of waste, are usually of lesser quality, and suffer from overuse. Once again, the answer is for the market to provide as much as possible to consumers and for people to take responsibility for their own choices, even when the costs are high.

On the subject of free stuff, have you noticed how more and more Americans cry out for government stimulus when the economy starts to slow down? Is this really a smart idea? And, if not, why? You know the drill—turn the page for answers.

CHAPTER DISCUSSION QUESTIONS

For a better understanding of the ideas in this chapter, work through these discussion questions in a small group:

1. Who pays the interest on "interest-free" student loans?
2. What are the two sources of income for the government?
3. When the government pays for free stuff out of the taxes it collects, who is really paying for it?
4. What is the downside of the government borrowing money to give its citizens free stuff? (Think interest.)
5. Explain this statement: "This truth—that all free stuff from the government must eventually be paid for from tax revenue—should make all of us concerned about how much our government gives away."
6. What are the short- and long-term effects of telling banks they cannot collect interest on student loans?
7. What was meant by the italicized statement at the end of this quote from the chapter: "In the case of student loan interest, helping the student by suspending interest payments on his loans hurts the banks, their employees, their suppliers, and consumers at large. *It is not a choice of suffering or not suffering; it is a choice of who suffers the most.*"
8. What "fair" rule was presented in the chapter for who should suffer when the economy declines?

9. Pick one of these problems with free stuff and explain it: the problem of waste, the problem of quality, or the problem of overuse.

10. The chapter presented the following "fix" to the problem of free stuff: "If you need the things the government is now giving you for free, you must begin to provide them for yourself. You must take responsibility for your own needs and stop looking to the government to provide everything for you." Is this a good or bad idea? Why or why not?

CHAPTER EIGHT: STIMULUS PLANS IN TIMES OF TROUBLE

I was born in the early 70s, and I do not even remember hearing the word "stimulus" until the 2000s. I received my first "stimulus check" sometime in that period from George W. Bush. He and Congress had passed a spending package that meant I got a check in the mail. As I recall, it was somewhere around $800. Ever since then, you can hardly go a year without some politician on TV talking about the need for "more stimulus." But what on earth are they talking about, and why does the government keep doing this?

UNDERSTANDING STIMULUS PLANS

We all want the economy to do well all the time. We like it when people have jobs, businesses make money (well, at least, some of us do), and the nation experiences financial and economic prosperity. But, as we all know, the economy doesn't always do so well. As outlined in chapter 1, economies can experience periods of recession and depression, where production, income, and employment decrease over time.

Many economists and politicians believe that certain actions by the government can affect the economy in a positive way. When a government makes a coordinated effort to improve the economy (to "stimulate" it, so to speak), we call it a "stimulus plan." As you know, these are all the rage these days, and most of us take for granted that they work (primarily because politicians tell us they do). But do they really? What does the evidence suggest? And, more importantly, are there any drawbacks to them?

parts of a stimulus plan

Great questions. Before we answer them, though, let's review the kinds of things included in stimulus plans. There are at least six.

STIMULUS PLAN DEFINED

A **stimulus plan** is an effort by the government to take a specific set of actions to help pull the economy out of a recession or depression.

Government spending: To create demand for products and services, the government will spend money on everything from roads and bridges to planes and bombs. This spending gives companies orders they otherwise would not get, keeping their employees employed and their factories running. This spending is intended to create a wave of activity the economy needs.

No matter how large the spending program is, however, it cannot by itself pull an economy out of the doldrums. The goal is to create a "spending multiplier," where each dollar the government spends is compounded down the line. Let me explain how this works.

If the government spends $100 million on new helicopters, for example, the firms that receive those orders turn around and spend some portion of that money on other things (buying parts, paying workers, and the like). And then the people who receive that income spend a portion of it, and on and on down the line until there is no money left to spend. The percentage that is spent is called the *marginal propensity to consume* (MPC). Don't let that term scare you. It's just a long-winded way of describing how much you spend of every dollar you receive. If you spend 100% of every dollar, the MPC is 1; if you spend 80%, the MPC is 0.8, and so on. The MPC is used to determine the total effect of a spending program on the economy (the "spending multiplier").

So, how much does $100 million in government spending actually produce in the economy? The amount depends on how much people spend of every dollar they receive, the MPC. The higher the MPC, the greater the impact on the economy. An MPC of 0.8, for example, results in a spending multiplier of 5, meaning if every person affected by the spending spent 80% of what they received, the $100 million would produce a $500 million impact on the economy.[43] If you only spend 50%, the multiplier would be 2, or $200 million.

FIGURE 8.1 - SPENDING TO STIMULATE

Government spending is a key part of stimulus plans.

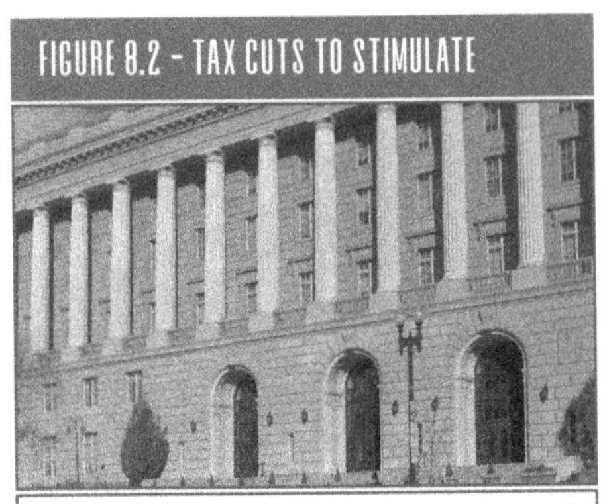

Tax cuts are used to stimulate the economy. Pictured here: The IRS Federal Building in Washington, DC.

The value of the multiplier varies by the type of spending and the economic situation at the time. It some extreme cases, it can even be negative. One source suggested that government spending programs can have multipliers of -0.3 to 3.6.[44]

The point here is not to overwhelm you with economic terms and calculations, but to show you the reasoning behind government spending in a time of decline. The government spends money during a recession to increase economic activity and reduce the severity and length of recessions. Does this actually work? We'll take up that question later in the chapter.

Cash payments to individuals or state governments: Another way to create demand is to give consumers and state governments money to spend. This is actually just another form of government spending, but because it is the consumer or state government that ends up making the purchases, we list it separately. The reasoning behind these kinds of cash payments is the same as the government spending discussed above—the multiplier effect. The money, it is assumed, will be spent, and the effect of that spending will ripple through the economy.

Tax cuts: Tax cuts are a way to put more money into the hands of individuals and businesses by reducing the tax rates that apply to their income. Tax cuts to individuals result in more money in their weekly paychecks (known as "discretionary income"), money they can spend on the things they need. Tax cuts to companies reduce the amount of tax they must pay to the government, giving them more money to invest in buildings, machines, and employees (known as, you guessed it, "investment spending"). Individual tax cuts are intended to create more demand; corporate tax cuts are intended to create more supply. Like gov-

---

43. Here is the calculation: First level players receive $100 million and spend $80 million (80% of $100 million). Second level players receive $80 million and spend $64 million (80% of $80 million). Third level players receive $64 million and spend $51.2 million (80% of $64 million). This continues until there is no money left. There is a formula for calculating what the total amount would be. I won't bore you with it here, but the result is $500 million.
44. Antony Davies, Bruce Yandle, Derek Thieme, and Robert Sarvis, "The U.S. Experience with Fiscal Stimulus: A Historical and Statistical Analysis of U.S. Fiscal Stimulus Activity, 1953-2011." *Mercatus Research*, April 2012.

ernment spending and cash payments to individuals, tax cuts are also subject to a spending multiplier, although it is usually smaller.

Cash payments or loans to businesses: To keep businesses from going under, or to help them keep their employees during slow economic times, the government will sometimes provide businesses with low interest loans or direct cash payments. In the case of loans, if certain terms are met, they may not have to be repaid. This was a central part of the COVID 19 stimulus plan of 2020. Cash payments and loans are also used as part of a bailout packages for large businesses, as in the case of General Motors in 2009.

Money supply adjustments: Through the Federal Reserve (known as "the Fed"), the nation's central bank, the government can make more money available for loans during a slow period in the economy. To understand how this works, you need to understand a little about how banks operate.

When you deposit money at a bank, the bank is free to use most of that money to earn a profit. Outside of special circumstances, the government requires banks to hold 10% of its deposits as reserves in case you, the depositor, come in and want your money back. The rest can be used to generate income, either by making loans or buying interest-bearing government bonds.[45]

Reserves are most often held at the Fed instead of the bank itself. This prevents the bank from having to hold a lot of cash in its vault. Vault cash must be secured (a costly endeavor), and it does not earn interest (you can't charge someone interest on money in your wallet). Cash held at the Fed, on the other hand, is extremely secure, and it can earn a small amount of interest.

How does money at the Fed earn interest? At the end of each day, if one bank is short on required reserves (the 10% it must hold for you), it can borrow that money from another bank holding money at the Fed. The interest rate charged to do this is called the "federal funds rate."[46] The federal funds rate becomes the starting point for the interest rates banks charge their customers because it is essentially their "cost" for money.

Now, back to how the Fed makes money available for loans. It does this in three ways. First, it can buy government bonds back from the banks. Remem-

THE FEDERAL RESERVE

The **Federal Reserve** is the central bank of the United States. Its job is to regulate banks, control the money supply, and maintain a healthy economy.

45. McEachern, *MacroEcon*, 252-256.
46. Ibid, 254.

TABLE 8.1 - FREEING UP MONEY FOR LOANS
1. The Fed makes more money available for loans by buying back its bonds from banks.
2. The Fed makes more money available for loans by reducing the rate for interbank loans.
3. The Fed makes more money available for loans by reducing the amount banks hold in reserve.

ber, as a rule, banks can only make money by making loans or buying government bonds. When the Fed wants banks to have more cash to loan, it will sometimes buy back its own bonds, thereby giving the bank more cash to loan out. A second way it creates more money is by reducing the federal funds rate, the rate banks get charged for loaning money to other banks. When the rate is lowered, banks can borrow money more cheaply from one another. A lower federal funds rate also means a lower cost of money for banks, which means they can offer lower interest rates to their customers. And, finally, the Fed can reduce the amount of reserves a bank is required to hold. It may drop the reserve requirement from 10% to 5%, for example. The bank is free, then, to loan the other 5% to customers.

Regulation adjustments: Businesses must meet thousands of government regulations to continue to operate. When the government removes or loosens these regulations, businesses can bring products to market faster and cheaper, and hire more employees, thereby growing the economy. Regulation changes can address environmental requirements, taxes, housing, food production, education, and transportation, among other things. Reducing regulation was a central part of President Donald Trump's administration. Although not part of a stimulus plan, per se, it was credited with helping drive the strong economy that existed during most of his administration (see Figure 8.3).[47]

FIGURE 8.3 - REGULATORY CHANGES

The reduction of business regulations was a key part of President Trump's economic plan in the late 2010s.

a brief history of stimulus plans

Although it may seem like a recent phenomenon, stimulus plans have been around for a long time in various forms. The first major

47. Diane Katz, "Here's how much red tape Trump has cut." *Heritage*, 17 October 2018.
48. Bill Dupor, "The Recovery Act of 2009 vs. FDR's New Deal: Which Was Bigger?" *Federal Reserve Bank of St. Louis*, 10 February 2017.

stimulus occurred during the Great Depression.[48] In 1933, the federal government enacted a number of programs designed to lift the country out of the worst economic crisis in its history. The New Deal, as it was called, used government agencies like the newly formed Works Progress Administration and the Agricultural Adjustment Administration to hire millions of employees for public works projects and to reduce farm production to increase prices. All told, the New Deal cost $41.7 billion at the time, over $811 billion in 2020 dollars.

FIGURE 8.4 - THE NEW DEAL

FDR used government stimulus to try to lift the nation out of the Great Depression in the 30s.

Later, the Eisenhower administration increased the money supply, lowered taxes, and eased down-payment requirements for home buying to fight the recession of 1953-54.[49] In response to the 1970-71 downturn, the Nixon administration deferred taxes on exports and altered exemption levels and credits on individual income taxes. President George H. W. Bush fought the 1990-91 recession with executive orders to reduce tax withholding and increase spending. His son, George W. Bush, to counter the downturn of 2001, made drastic tax cuts and initiated a $36 billion tax rebate program, and, as a first shot at stemming the economic effects of the Great Recession of 2007-09, enacted a $113 billion tax rebate and business tax reduction program. President Obama followed with his own set of stimulus actions in 2009, including tax reductions and spending initiatives totaling $787 billion.

All of these, however, pale in comparison to the stimulus passed under President Donald Trump's administration in response to the Coronavirus pandemic. The Coronavirus Aid, Relief, and Economic Security (CARES) Act included, among other things, funds for additional unemployment pay, loans to businesses to maintain workforce levels, direct payments to consumers, and financial support for local and state governments. The initial cost estimate for the CARES Act was more than $2 trillion.[50]

DO STIMULUS PLANS WORK?

As the brief review above shows, stimulus plans have been used extensively over the last 90 years and have cost trillions of dollars. But what do we have to

49. Davies, Yandle, Thieme, and Sarvis, "The U.S. Experience with Fiscal Stimulus."
50. U.S. Department of the Treasury, "The CARES Act Works for All Americans."

FIGURE 8.5 - THE BIGGEST STIMULUS TO DATE

President Trump and Congress passed the $2+ trillion CARES Act in 2020 in response to COVID-19.

show for all this money and effort? Is there any evidence these plans actually worked?

A 2012 report issued by the Mercatus Center at George Mason University reviewed the impact of stimulus plans used by the American government from 1953 to 2011.[51] This effort included a review of the Economic Report of the President (ERP) for every administration responding to a downturn in that period (an ERP is a document issued by the president detailing the status of the economy and any actions his administration has taken to address economic issues), and then evaluated the actual effect of those plans on the economy. What did it find?

Overall, the report concluded that stimulus plans have little to no effect on the growth of the economy. Here's how it summed up their impact: "There is no significant relationship between changes in government spending during recessions and economic growth at any point from one to 10 years later. In summary, regardless of the time horizon and regardless of whether the economy is in recession or expansion, the data exhibit no evidence of stimulus spending having any effect on economic growth." In other words, in spite of claims to the contrary, government stimulus action does not bring the economy out of recession.

But why?

The Mercatus report noted several reasons. First, there is the *recognition problem*. Governments have a difficult time recognizing when a recession has actually started and determining the factors causing it. Most often, these facts are only understood in hindsight. If you don't know you are in a recession, or if you don't know why you are in a recession, it is difficult to do anything about it. Second, there is the *timing problem*. Measures to correct the economy need to be taken at the right time to address the issues when they are happening. Governments find it difficult to respond in a timely fashion to rapidly changing economic conditions. And, when they do respond, they oftentimes continue acting after the problem has passed, causing more harm than good. Then

STIMULUS PLAN RESULTS

According to the Mercatus report, stimulus plans enacted since WWII have had no noticeable effect on the growth of the economy following their usage.

51. Davies, Yandle, Thieme, and Sarvis, "The U.S. Experience with Fiscal Stimulus."

TABLE 8.2 - WHY STIMULUS PLANS FAIL
1. The government is slow to recognize a downturn and to understand its true causes.
2. The government tends to respond too late and apply its responses for too long.
3. The spending multiplier is too hard to predict and not as high as the government would like.

there is the *coordination problem.* The stimulus plan elements must be carried out in concert with one another. The size and nature of government usually prevents this from happening.

Another issue the Mercatus report mentions is the *multiplier problem.* As stated earlier, the goal of government spending during a recession is to create a domino effect in the economy. The hope is that a small spending program (in terms of the whole economy) will compound to produce a large impact through the spending multiplier. But multipliers only work if the people who receive the money actually spend it. During tight economic times, individuals often use cash payments and tax credits to pay down debt or increase savings. State governments do the same. A study by Valerie Ramey found that multipliers for stimulus programs can range from -0.3 to 3.6 (a negative multiplier means that the economy actually got worse after the stimulus spending). She also found that state governments had varied levels of spending after receiving federal stimulus dollars, showing multipliers of -0.6 to 2.0.[52]

BIGGER PROBLEMS WITH STIMULUS PLANS

The fact that stimulus plans don't work very well is not their only weakness. They also create other issues that can lead to greater problems in the future. The most important of which is the debt they cause.

stimulus plans and debt

The one thing that never seems to enter the discussion when stimulus plans are being considered is the fact that they are all funded by debt. The U.S. government has not run a surplus in its yearly budget since 2001. Since then, every program above normal spending has been paid for by borrowing (and much of normal spending has been paid for this way as well). This means that the stimulus plans of 2001, 2008, 2009, and 2020 have all been paid for *by borrowing money from others.* In that period alone, the federal debt has risen from around $6 trillion in 2001 to over $27 trillion at the end of 2020, an increase of over $21 trillion or 450%

52. Ibid.

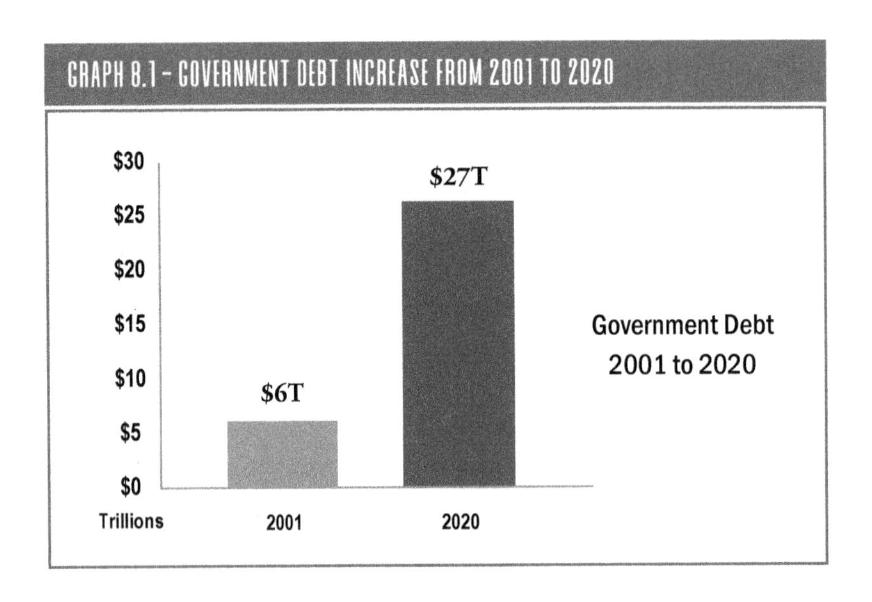

(see Graph 8.1). Not all of this is stimulus related, of course, but it illustrates the truth that stimulus plans are 100% debt. The Mercatus report states the case very nicely: "We find that a long-run price must be paid for short-run stimulus. Indeed, the long effects swamp the short-run gains. There is no such thing as a free stimulus."[53]

the larger spiritual issues revealed by stimulus plans

Stimulus plans are a result of the people's expectation that government should always do something to reverse hard times. If the people didn't expect these actions from government, politicians would be much less likely to take them. So, why do Americans demand stimulus plans, and what does that reveal about our society overall?

The demand for stimulus plans shows an increasing desire among Americans to avoid any kind of suffering. We have gotten used to things going well all the time and refuse to allow them to get tough. We have become such a pampered, soft society that the idea of having to struggle or endure any kind of hardship, economic or otherwise, is absolutely unacceptable to us. So, when the economy seems to be heading south, we cry, "Government, do something! We can't bear hard times!"

53. Ibid.
54. See Appendix B for a review of why suffering exists and how God uses it.

God chooses to put individuals and nations through hard times to bring about a greater good in their lives.[54] Due to our nature as human beings, we are prone to stray and make decisions that will destroy us. Hard times are not always divine punishment (though they certainly can be); sometimes they are divine protection.

WHY AMERICA IS SUFFERING

America is suffering because it has moved away from the Creator's rules and design for social, moral, and economic life. Our sin has caused our suffering.

The author of Psalm 119 understood these truths. First, he saw that the trial he had endured was given by God out of love for him. He writes, "I know, O LORD, that Your judgments are right, and that in faithfulness You have afflicted me" (75). He then saw how his difficulty taught him the ways of God. "Before I was afflicted I went astray," he says, "but now I keep Your word" (67). And finally, he realized that, for these reasons, trial was actually good for him. "It is good for me that I have been afflicted, that I may learn Your statutes," he concluded (71).

America is suffering, economically and otherwise, because it has moved away from the Creator's rules and design for life. We like to think of ourselves as somehow liberated from God's prison when we throw off the teachings of the Bible. But we only make things worse for ourselves when we do. The suffering God has been trying to bring us through difficult economic times has been sidestepped, at least for a season, by debt. The pain that these trials could have brought, and the changes they could have worked in us, would have made the future brighter. Instead, because of our foolishness, the future is bleaker than ever, as I will show later in this book.

Another truth that is revealed through our near addiction to stimulus plans is our growing trust in government to deliver us. In our time of need, we do not turn to the God of the Bible anymore in America. Our first call is not to prayer to our Creator, but supplication to our legislators. We have adopted the errant view that because the government is so big, and has access to so much money and resources, that it can do anything—stop a recession, overcome a storm, or cure a plague.

TABLE 8.3 - SPIRITUAL ISSUES & STIMULUS ADDICTION
1. Our appetite for stimulus plans reveals our desire to avoid any and all suffering.
2. Our desire for stimulus plans reveals our growing trust in government to deliver us.
3. Our desire for stimulus plans reveals our lack of individual preparation for times of trouble.

This is also where the connection between socialism and stimulus plans can be seen. As indicated in other places in this book, a key goal of socialism is to gain control over the population. One of the ways to do that is to make the people dependent on the government for everything. Free government programs, as discussed in chapter 7, are one way to do this. Another way is to be the one everybody turns to in a crisis, and this is where stimulus plans come in. The government provides them to "rescue" the helpless masses, and, in return, the people give over more of their rights to the government.

In spite of the trust many have in the federal government, we are on a clear path to learning its limits. As I will discuss in detail in the next chapter, the federal government is on its way to a complete financial collapse that will bring the whole world down with it. If we do not change course soon, our all-powerful Uncle Sam will not even be able to help himself, much less the rest of us. God allows no competition for his throne, and he is about to put down the idol that is the American government in a spectacular way.

Finally, our reliance on government support in time of need has caused us to forgo the responsibility to take care of ourselves during hard times. Instead of saving for a rainy day, we wait for a stimulus check; instead of buying extra food to get through an emergency, we wait for Uncle Sam to deliver what we need to our doorstep. We have forgotten how to be wise about the future, and that lost knowledge will burn us severely in the not-too-distant future.

BUT THE DEBT!

Stimulus plans, as well-intentioned as they may be, make virtually no difference in the economy. All they do is add to our long-term debt, prevent us from learning the lessons God has for us in national trials, give more of our freedoms over to the government, and place our nation in greater jeopardy in the future. Trusting in them, therefore, and the government behind them, is a useless and ultimately doomed enterprise.

The individual threads have now been stretched out and are ready to come together in a fine, sharp, and devastating point. How do the socialist ideas and debt-ridden activities of the last eight chapters come together? What impact will they have if we continue on this course? Chapter 9 will explore these questions and more.

CHAPTER DISCUSSION QUESTIONS

For a better understanding of the ideas in this chapter, work through these discussion questions in a small group:

1. What is a "stimulus plan"?
2. The chapter discusses six elements that can be used in stimulus plans. Which did you find most interesting? Why?
3. What is the "spending multiplier"? Why is it so important to the idea of stimulus plans?
4. The chapter outlines three ways the Fed makes more money available for loans. Explain any one of them.
5. Were you surprised to see how often stimulus plans have been used in America since the Great Depression? Why or why not?
6. What did the Mercatus report conclude about whether or not stimulus plans actually worked? What were the four reasons it gave for this result?
7. The Mercatus report discussed research by Valerie Ramey on the spending multipliers for stimulus programs. What range of multipliers did she find for stimulus plans in general? (It starts with -0.3.) What does this tell us about the effectiveness of stimulus plans?
8. Give the dollar amount and percentage increase in the U.S. national debt since 2001. Are you surprised by this increase?
9. Why does God send nations through difficult economic times? What is he trying to do through these situations? Is this good or bad for us?
10. What does the reliance on government support in time of need do to our own individual efforts to take care of ourselves in difficult times? Is this a good or bad thing? Why?

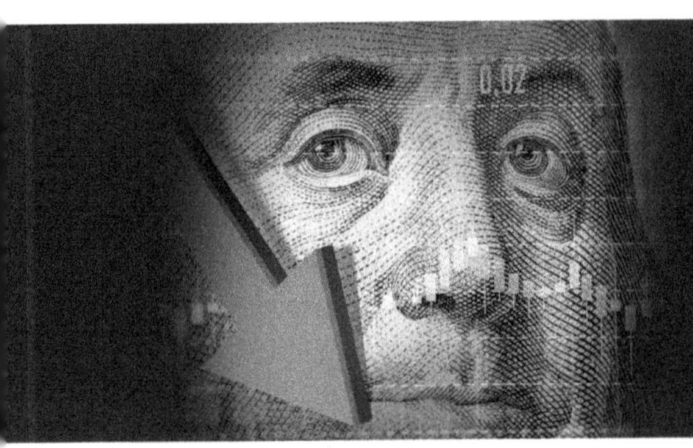

CHAPTER NINE: THE TIME BOMB THAT IS THE NATIONAL DEBT

The issues addressed in the previous eight chapters come together here. The socialist ideas taking hold in the nation and the growth of government spending and deficits are not independent issues. Together, they combine to create a single, overwhelming threat to the nation. That threat is the national debt.

While many may not understand the significance of our debt, most are aware that it exists, and that it is massive. The debt is like that belligerent, drunk uncle who comes to the family Christmas party every year. Everybody knows he's there, and everybody knows he could blow up at any moment, but instead of doing anything about him, the whole family just ignores him and prays he will behave. Well, it's time to take a closer look at this problematic uncle of the American people. It's time to come face-to-face with Uncle Sam's debt.

THE BIG PICTURE OF OUR DEBT

> **DEFICIT VS DEBT**
>
> A **deficit** is a yearly shortage in the government's budget. The total amount of yearly deficits creates the **national debt**, the sum of what the nation owes.

The national debt is the total outstanding amount owed by the federal government at a certain point in time. It is the amount it has borrowed but has yet to pay back. As of the end of 2020, that amount was $27 trillion.

The national debt is made up of the government's yearly deficits. A deficit is the difference between what the government brings in (taxes) and what it pays out (spending) on a yearly basis. The government's accounting year is a bit confusing because it runs from October 1 of one year to September 30 of the next year. This is called the "fiscal budget year."[55] To avoid the confusion associated with this, we will use numbers for the calendar year in this chapter. The deficit for the 2020 calendar year through the end of October was $3.7 trillion.[56]

It is also important to point out the amount of interest paid on the debt. In the last 10 years alone (2011-2020), the Federal government paid $4.9 trillion in interest on its debt. That is enough to cover more than 18 months of the government's income. In other words, we have spent more than a year and a half worth of income on interest in the last 10 years, but we have paid nothing towards the principal (the amount we actually borrowed). This means nearly 40% of our debt in the last 10 years *has been interest we pay on that debt.* We are now in the unthinkable position of borrowing to pay interest. Table 9.1 shows the yearly deficits for each year from 2011 to 2020 and the total interest paid.[57]

TABLE 9.1 – U.S. DEFICITS & INTEREST

Year	Budget Deficit (Trillions)	Interest on Debt (Trillions)
2011	$1.3	$0.423
2012	$0.9	$0.316
2013	$1.0	$0.439
2014	$0.5	$0.429
2015	$1.2	$0.454
2016	$0.6	$0.476
2017	$1.3	$0.541
2018	$1.1	$0.582
2019	$0.7	$0.582
2020	$3.7	$0.547
Totals	$12.3T	$ 4.9T

HOW THE GOVERNMENT BORROWS MONEY

The government goes into debt by offering bonds for sale through the U.S. Treasury Department.[58] Bonds are a type of debt instrument used by governments to raise money.[59] When you buy a bond,

55. The word "fiscal" comes from the Latin word *fiscālis*, which refers to matters involving government finances or spending. Dictionary.com app.
56. Federal Reserve Bank of St. Louis, "Federal Debt: Total Public Debt."
57. Federal Reserve Bank of St. Louis, "Federal Surplus or Deficit."
58. There are several types of bonds the Treasury issues. The primary difference is the length of time before they "mature" and the principal of the bond (the amount paid for it) is repaid.

FIGURE 9.1 - AN INSTRUMENT OF DEBT

Example of a $1,000 Series EE Savings bond.

you are essentially loaning money to the government. Your reward for doing so is the interest the government will pay you for using your money. This interest is often paid when the bond "matures," or comes due, but it can be paid during the year as well.

U.S. bonds are issued for different amounts of time, ranging from four weeks to 30 years. The Treasury sells bonds through auctions. The interest rates paid on the various types of bonds are determined during these auctions.[60]

The total debt of the federal government is often broken down into two components: the debt the government owes to itself and the debt the government owes to the public. The debt the government owes to itself, what I call "internal debt," is what the U.S. Treasury owes to other parts of the government. Departments which have excess funds sometimes take those funds and buy bonds from the Treasury to earn interest. The Social Security Trust Fund, for example, uses the excess income it receives from social security payroll taxes to buy government bonds. This allows it to receive a small return on the money it is holding. When this happens, the Treasury incurs a debt to the Social Security Trust Fund. This is an example of internal debt because it is debt that must be paid back to another part of the government, not a third party. As of the end of 2020, the total amount held for all government agencies was about $6 trillion, or about 22% of the government's total debt.[61]

Debt sold to those outside of the government is referred to as "debt held by the public." This debt includes that purchased by U.S. buyers, like individuals, banks, and mutual funds, as well as overseas buyers, like private investors and central banks of foreign nations such as China and Japan.[62] At the end of 2020, total debt held by the public was around $21 trillion, or about 78% of the total debt.

The reason this distinction is important is that the government sometimes plays games with how it reports its debts and the interest it pays on those debts.

---

59. Bonds are also used by local and state governments and corporations.
60. Investopedia, "U.S. Bonds vs. Bills and Notes: What's the Difference?" 24 July 2020.
61. The amount of U.S. debt held by the Social Security Trust Fund is $2.9 trillion. Kimberly Amadeo and Somer G. Anderson, "Who Owns the US National Debt?" *The Balance*, 14 October 2020.
62. Congressional Budget Office, "Federal Debt: A Primer." March 2020.

Many times, politicians will talk about "public debt" instead of "total debt." They do this because, as we have seen, public debt is lower than total debt. In fact, the government, when reporting its spending for a prior year or its budget for an upcoming year, only reports the interest it must pay on public debt, not the total debt. For instance, the Congressional Budget Office (CBO) reported that interest payments were only $375 billion in 2019[63], but the Federal Reserve reported the total at $582 billion.[64]

> **INTERNAL VS PUBLIC DEBT**
>
> **Internal debt** is what the government owes to its departments. **Public debt** is what is owed to the rest of the world. **Total debt** is the combination of the two.

Why the difference? If you look closely at the CBO report, you will see that it refers to "government interest payments on debt held by the public." The problem is that this is not an accurate measure of the interest that has been paid on the government's debt. In fact, it is understating the interest by nearly 36%. This makes regular citizens like you and me think the impact of the debt is less than it is and may lead us to support further borrowing when we otherwise would not.

SAMMY SPENDER: THE NATIONAL DEBT AS AN INDIVIDUAL

The debt numbers are so big it is hard to get our minds around them. Like studying the size of the universe, at some point we lose the ability to keep the figures in perspective. How does the average Joe determine if a $27 trillion dollar debt is really that bad, especially given the size of the U.S. economy? One way to get a better sense of the scale of the debt is to pretend the U.S. government is a single individual. We'll call him "Sammy Spender." Now, let's ask this question of Sammy, "If Sammy made $100,000 per year and had the same spending habits and debt levels as the U.S. government, what kind of financial shape would he be in?" Let's do that now.

If Sammy were the U.S. government, he would spend an average of $134,000 every year, even though he makes only $100,000. He would borrow the difference—$34,000—from his family, friends, and even his enemies. His total debt from credit cards, car loans, and mortgages would be a whopping $900,000. Each year, he pays over $18,000 in interest on that debt ($1,500 a month) but not a cent on the principal itself (in other words, the total he owes never goes down). At his present pace, his debt will be $1,222,000 in 10 years, and his interest payments

63. Congressional Budget Office, "The Federal Budget in 2019: An Infographic." 15 April 2020.
64. Federal Reserve Bank of St. Louis, "Federal Surplus or Deficit."

TABLE 9.2 - SAMMY SPENDER'S FINANCIALS	
Yearly Income	$100,000
Yearly Spending	$134,000
Current Debt Level	$900,000
Debt-to-Income Ratio	900%
Yearly Interest Payments	$18,000
Debt Level in 10 Years	$1.22M
Debt-to-Income Ratio in 10 Years	1222%
Interest Payments in 10 Years	$24,000

will be almost $24,000 per year (about $2,000 a month). And, if interest rates go up even slightly, he could pay a lot more.

To make things worse, last year Sammy borrowed $106,000 to help his friends out of a jam (the first 2020 COVID stimulus), and he is borrowing another $30,000 right now because the original amount wasn't enough (second COVID stimulus). He is also thinking about borrowing the money to pay off his friend's student loans, a gesture that will cost him another $63,000.

Now, imagine you are a financial planner and Sammy comes to you with these figures. What would you tell him about his financial future? He has a $100,000 income, $900,000 in debt, and is borrowing money left and right, not only to finance his own lifestyle but also to take care of his friends. How long before this man goes completely bankrupt and is sleeping on the street? It doesn't take a financial genius to see that he is headed for a train wreck of epic proportions. His entire life is about to be destroyed by this debt. There is simply no other conclusion that can be drawn.

This is the state of the U.S. government—*your* government—right now. Frightening doesn't even scratch the surface, does it? But wait, it gets worse.

HOW WE EXPLAIN AWAY THE DEBT

Politicians and economists find clever ways to gloss over the numbers you just read. Their creativity in making us believe our debt is no big deal seems to have no limits. A good example of this is found in the textbook *Macroeconomics* by Dornbusch, Fischer, and Startz published in 2018 by McGraw-Hill Education, one of the largest educational publishers in the country. This is actually the text used in my graduate-level macroeconomics course at Liberty University. Let's consider the arguments these authors' make and a rebuttal of each.

The first argument used to make us believe U.S. debt is relatively low is to compare it to the size of the total economy (the gross domestic product, GDP).

The debt level is divided by the total GDP for that year to determine a percentage. If the percentage is below 100%, the debt level is deemed acceptable. The 100% level has no statistical significance, and economists debate whether this percentage is actually meaningful. However, the thinking may be that once the debt goes above 100%, the nation owes more than it can earn in one year (GDP, as noted in chapter 1, is also considered the income level of the nation). In this sense, the debt-to-GDP ratio is really a debt-to-income ratio for the entire nation.

TABLE 9.3 - U.S. DEBT-TO-GDP %

Year	Debt-to-GDP %
2011	93%
2012	97%
2013	101%
2014	103%
2015	101%
2016	104%
2017	103%
2018	104%
2019	104%
2020	135%

There are a couple of problems with this measure. First of all, the debt-to-GDP ratio has been above 100% since 2013. In fact, it hit 107% at the beginning of 2020 and 135% by the end of the second quarter. (See Table 9.3 for yearly debt-to-GDP percentages.) So, if you're going to be bothered by a 100% plus debt-to-GDP ratio, you've had since 2013 to do it.

A second problem with the 100% debt-to-GDP ratio is that it has nothing to do with the government's ability to repay its debt. The government does not have access to the full GDP to use to pay its creditors. That money is in the wallets of the American people. This would be like evaluating Sammy's ability to pay back his debt by counting his brothers', sisters', parents', aunts', uncles', and all his friends' income along with his. Sammy can't take all their money to pay back his loans, so he can't count it when evaluating his debt level.

In the same way, the government only has access to *its* income to pay back its debts, which, in this case, is its tax receipts. Over the last 10 years, the government has averaged a little over $3 trillion in total tax revenue each year. Using this as a basis for calculation, the government's debt-to-income ratio is not 135%—it is 900% ($27 trillion in debt / $3 trillion in annual income).

The Dornbusch, Fischer, and Startz textbook also suggests that the portion of the debt held by government entities, like the Social Security Trust Fund[65], is not as much a concern because it is essentially the government "owing itself."

65. Others include the Office of Personnel Management Retirement, the Military Retirement Fund, and Medicare. Amadeo and Anderson, "Who Owns the US National Debt?"

GOVERNMENT REPAYMENT

The government can only use its income to repay its debts, not the wealth of all Americans. Its income is only what it can raise in taxes, which is limited.

Furthermore, the elements of the debt held by the U.S. public, including U.S. citizens and organizations, state and local governments, and individual and corporate investors, are really "owed to ourselves," and are, therefore, not as significant as that which is owed to international investors or foreign governments. When you take out the money "owed to ourselves," it is argued, the debt becomes much smaller.

This view is flawed for many reasons. In the first place, since the federal government is one entity, how can it actually "borrow" from itself? How could I, as an individual, borrow from myself? To do this, I would have to take money from one account I own and move it to another account I own. In that case, I may be "borrowing," in a sense, but in another sense, all I am doing is playing with numbers. I still have the same total amount of money, and I still have the same total amount of debt.

Even allowing for the government to borrow from itself, those debts between government agencies are still real debts. The money represented by those debts is money the American people have given over to the government. It is relied on for specific purposes down the road and therefore must be repaid. Take the Social Security Trust Fund, for example. When you retire, are you not fully expecting the government to pay your benefits? Of course, you are. The government "owing itself" doesn't change the fact that one day it must pony up and pay the benefits that are represented by those debts.

In the same way, funds owed to U.S. citizens and organizations, state and local governments, and private investors cannot just be written off. Even if we "owe it to ourselves," in the sense that the American government owes the American people, we still *owe it*. Individuals, state and local governments, and companies are counting on the principal and interest from those debts for their livelihoods. And what would happen if the federal government decided not to pay back those who constituted "itself"? Would we all just forgive the debt and walk away? Or would our companies, retirement accounts, and state and local governments be irreparably harmed? The consequences would be disastrous.

FIGURE 9.2 - LOANING OUT SOCIAL SECURITY

The Treasury has borrowed all extra funds from Social Security and is obligated to pay them back with interest.

A final argument this text makes is that, because a national government is a sovereign, it can simply default on its debts, especially those owed to foreign nations. It states:

> Because the government is the "sovereign"—the maker of rules—the government can just decide to not pay off its debts. In principle, there is little or nothing other countries can do when faced with a sovereign default other than to refuse to lend to that country in the future. Since being shut out of international markets can be a steep price, there is a large incentive for countries not to default. At the same time, there is often immense internal political pressure to default on debts owed largely to foreigners rather than cut domestic spending.

The idea here is that the government can do anything it wants because no one has the power to stop it. This is bad reasoning on at least two counts. In the first place, who is to say a foreign power might not go to war with us because of debt? At the end of 2020, the Chinese government owned more than $1 trillion worth of U.S. bonds.[66] With tensions between the U.S. and China rising in recent years, and with the growing power of the Chinese military, a physical conflict over debt is not out of the question. (See Table 9.4 for a list of top foreign owners of U.S. debt.)

Another thing to keep in mind here is the impact of a foreign debt default on stateside investors. When someone defaults on their mortgage, are other lenders more or less likely to loan them money, even if they themselves have been paid on time? Less likely, of course. When we default in one area of our debt, that sends a signal to the rest of the market that we are having serious finan-

TABLE 9.4 - U.S. DEBT HELD BY SELECT NATIONS

Nation	U.S. Debt Held/Trillions
Japan	$1.3
China	$1.1
United Kingdom	$0.430
Brazil	$0.316
Luxembourg	$0.263
Switzerland	$0.255
Hong Kong	$0.246
Total by All Nations	$7.1

66. Weizhen Tan, "The growing U.S. deficit raises questions about funding as China cuts U.S. debt holdings." *CNBC*, 2 November 2020.

cial issues and may default in other areas, including to them. As a response, lenders will not extend us any further credit until they are sure we have regained our financial footing. It would be no different with U.S.-based lenders and the U.S. government.[67]

Even if the U.S. hadn't defaulted on debts with U.S.-based lenders, those lenders would be fools to buy any more U.S. debt for fear that they would be the next ones to be defaulted upon. This would cause U.S. credit to dry up almost overnight, the federal government would not be able to borrow enough money to meet its obligations, and what it could borrow would be at significantly higher interest rates. A default to foreign creditors would bring the U.S. government and economy to its knees.

WARNINGS FROM THE FOUNDERS AND GOD ABOUT DEBT

While debt may seem as American as apple pie, the Founders of our nation constantly warned us against its charms. Thomas Jefferson, in a letter to Samuel Kercheval, said, "To preserve [the people's] independence, we must not let our rulers load us with perpetual debt."[68] Here, he ties the freedom of the people to the level of debt of their government. Keep this in mind when we discuss the Bible and debt later in this section.

In a report while secretary of the treasury, Alexander Hamilton discussed his dislike for debt by saying, "Nothing can more interest the national credit and prosperity, than a constant and systematic attention to...extinguishing the present debt, and to avoid, as much as possible, the incurring of any new debt."[69] Hamilton saw large debts as corrosive to both the credit worthiness of the nation and its welfare, points we have discussed previously in this book.

FIGURE 9.3 – FOUNDERS' WARNINGS

Alexander Hamilton and other Founders warned the nation against too much debt.

John Adams, the second president of the United States, had this to offer regarding the nation and debt, "The consequences arising from the continual accumulation of public debts in other countries ought to admonish us to be careful to prevent their

67. This is rational self-interest at play again.
68. The Jefferson Monticello, "Extract from Thomas Jefferson to 'Henry Tompkinson' (Samuel Kercheval)."

growth in our own. The national defense must be provided for as well as the support of Government; but both should be accomplished as much as possible by immediate taxes, and as little as possible by loans."[70] To Adams, all spending programs, even that of national defense, should rely on the government's income, that is, it's tax revenue, not on its ability to borrow.

To top off the conviction of our Founding Fathers on the subject of debt, let me quote George Washington, our first president, from his final address to the nation[71]:

> As a very important source of strength and security, cherish public credit. One method of preserving it is to use it as sparingly as possible, avoiding occasions of expense by cultivating peace, but remembering also that timely disbursements to prepare for danger frequently prevent much greater disbursements to repel it; avoiding likewise the accumulation of debt, not only by shunning occasions of expense, but by vigorous exertions in time of peace to discharge the debts which unavoidable wars may have occasioned, not ungenerously throwing upon posterity the burden which we ourselves ought to bear.

Washington viewed low debt levels as a source of strength and security for America. To prevent significant borrowing, he recommended the nation stay out of wars, limit spending, and aggressively pay off debts once they were accumulated. It seems we have refused his wisdom on all counts.

The Founding Fathers were not the only ones who warned us of debt. God himself, in Proverbs 22:7, said, "The borrower is servant to the lender," and in Romans 13:8, he advised, "Owe no one anything except to love one another." The law of debt, as I will call it, that Proverbs 22:7 outlines is a universal truth in human life: If you do not manage your debts, you will eventually become the slave of the party or parties from which you borrowed. The debt will begin to rule your life and, if you are not careful, ruin it.

THE LAW OF DEBT

The **Law of Debt** is the truth that debt cannot be taken out forever. Eventually it must repaid, and, if too much is borrowed, it will destroy the borrower.

69. National Archives, "Fact No. 1, [11 September 1792]."
70. Yale Law School, "First Annual Message of John Adams."
71. Senate.gov, "Washington's Farewell Address to the People of the United States."

As I consider the wisdom of the Founders and the Creator, I cannot help but think that America is somehow in a debt trance. Our debt levels get bigger and bigger with every passing year, but we cannot seem to grasp where all of this is headed. We keep finding new things to borrow for every day. The slavery to this debt is coming, and, in fact, is already here, but we are completely blind to it.

I am reminded of the old question: How do you cook a frog? You don't drop him in a boiling pot, they say. You put him in lukewarm water and slowly turn up the heat. Because of the gradual change, he doesn't even realize he's being cooked.

America is a lot like that where debt is concerned. The water all around us is scalding to the touch, while we sit snuggly in our spot, wide grins on our faces. The difference between us and the frog, however, is that there isn't a third party turning up the heat. No enemy or foreign government is boiling us alive. We are doing it to ourselves. Instead of someone else reaching over to ratchet up the temperature, we reach over ourselves, the boiling water bubbling around us, to edge the burner up a little farther. And every time we approve another stimulus or another bailout or another free-whatever-plan, we bring ourselves one step closer to burning alive.

A SIMPLE PLAN FOR REPAYING THE DEBT

I hope you see the dangerous position we are in as a nation because of our debt. It is by the grace of God alone that our foolish financial decisions have not caught up to us yet. It is still possible to make the changes necessary to avoid a future economic catastrophe.

At the end of the day, the solution to our problem is not that complicated: We must get out of debt. But how do we do this? It's actually very simple. Just ask yourself, "If I was in this kind of debt, how would I get out of it without going bankrupt?" There really is only one way, right? Increase your income, cut your expenses, and pay all your extra money on your debts. Depending on how large your debts are, it may take a while, but it is possible, isn't it? Sure, it is.

It's no different with the U.S. government. It has a debt it must repay. The only way to do it is to do what you and I would do—cut expenses, increase income, and pay down the debt.

Ultimately, however, our goal should not just be to get out of debt; it should also be to become a nation that saves. We will need funds in the future to respond to emergencies and accomplish goals above and beyond our budget. Therefore, getting out of debt is not enough. We must also create a situation where we can save what we need for the future *so we never have to borrow again.*

How can this be accomplished? Let me offer a simple, four-pronged plan.

1. ***Balance the budget over 10 years.*** Cutting spending and raising taxes too much too fast could be devastating to the economy (I will explain why in the next chapter). Therefore, we need to do this in smaller increments. At present levels of spending and income, reducing spending and increasing revenue by $500 billion each over 10 years will get the budget balanced.
2. ***Create a budget surplus of $500 billion over the following 10 years.*** In other words, continue to reduce spending and increase taxes so that, at the end of the following 10 years (20 years total), the net surplus for the budget is $500 billion per year. Whatever the surplus is in each of the years until that point, apply it to the debt.
3. ***Maintain a $500 billion surplus each year until the debt is paid off.*** Depending on what debt levels are by the time the budget is balanced and the surplus is attained, the debt could be paid off in 70 to 80 years.
4. ***Build savings of one year of GDP over the next 20 to 30 years.*** Use the budget surplus, plus the additional funds freed up in interest payments, to create a national reserve fund to be used in emergencies.

Yes, this plan is very simple, and, yes, it relies on a number of assumptions, not the least of which is no major economic, political, or military crises while it is being carried out. Moreover, the Federal budget and tax code are tremendously complex, and many details would have to be worked out to implement such a plan. However, the math is undeniable. If we can get to a point of surplus, we can pay off our debts and begin saving in 100 years or less. And, within 120 years or so, we could have a years' worth of income saved.[72] We would no longer need to borrow from anyone.

THE DIM PROSPECTS FOR REPAYING THE DEBT

Unfortunately, implementing this simple solution is not so easy in the modern American political and social climate. There are at least three major roadblocks that may prevent it. Let's consider those now.

1. The pressure to increase spending. While we may recognize the need to reduce spending, in reality there are both legal and social pressures that almost

TABLE 9.5 - PLAN FOR DEBT REPAYMENT
1. Balance the budget over the next 10 years.
2. Create a budget surplus of $500 billion over the following 10 years.
3. Maintain a $500 billion surplus each year until the debt is paid off.
4. Build savings of one year of GDP over the next 20 to 30 years.

guarantee it will grow drastically instead. The Social Security Act, for example, requires that retirees are paid retirement benefits after a certain age. This is not optional; it is demanded by law.

The problem is that social security obligations in the future will far outweigh the money collected by the program. And, if you'll recall, the excess brought in thus far for social security has all been loaned to the Treasury and spent. A current estimate for the cost of social security over the next 75 years is around $14 trillion.[73] Unless the law changes, spending for legally mandated retirement and medical programs like this will drastically add to government spending.

On top of this, more and more Americans are calling for free government handouts in all forms. From student loan forgiveness to lucrative unemployment benefits to free medical care, the American public wants more from its government, not less. This trend will make it extremely difficult for the government to trim its spending.

2. ***The limited potential of revenue increases.*** While spending projections are on the rise, the ability for the government to increase its income is restricted. Calls for higher taxes on rich individuals and corporations grab media headlines, but history shows that higher tax rates for these groups do not necessarily translate into higher revenue for the government.

The Mercatus report referenced earlier evaluated the results of revenue increases through higher taxes.[74] The report noted that average tax rates have changed several times over the last 70 years, but tax revenue has stayed virtually the same. Between 1950 and 2006, tax rates had a low of around 26% and a high of about 34%. Tax revenue during that time, however, was only 0.7% higher under the highest rate than the lowest. In other words, even when taxes were raised by

72. This may seem like a long time, but keep in mind, it took 75 years to get to this level of debt. We need to be a people with a long-term outlook, not the kind that looks for a quick fix for everything. Even though the ultimate goal is far away, we will begin to reap the benefits in a much shorter time.
73. Kathleen Romig, "What the 2019 Trustees' Report Shows About Social Security." *Center on Budget and Policy Priorities*, 5 June 2019.
74. Davies, Yandle, Thieme, and Sarvis, "The U.S. Experience with Fiscal Stimulus."

nearly a third, tax revenue didn't rise even 1%. Federal tax income during that entire period averaged 18% of GDP, even with different tax rates.

The reason for this lies in the issue of incentives discussed earlier in this book. At some point, higher taxes in free societies become such a disincentive that people do not work the extra hours or take the additional risk associated with creating income that can be taxed. This tendency is known as the "Laffer curve." In addition, when taxes are raised on the wealthy, lawmakers often create loopholes that can be used to dodge some or all of the additional tax burden (which includes themselves, by the way—most congressmen and women are millionaires). The only income levels where higher taxes actually produced higher revenues were in the middle and lower classes, groups most politicians are afraid to go after.

Not only is significant revenue difficult to acquire through taxation, but America's move toward socialism is likely to reduce its wealth substantially over time, thereby limiting the government's pool of funds from which to draw taxes. Socialist ideas like high taxes on the rich, high minimum wages, increased government control over portions of the economy, and increasingly expensive social programs will slowly weigh down the economy's ability to produce wealth. As this happens, government income will decrease even further.

3. The lack of political and social will. The changes needed to reduce the federal debt will bring difficulty to the American people. We will all have to pay higher taxes (to some extent, at least), go with less government programs, and do without the things we now consider necessities. But a growing number of Americans will have none of this. We have lived "high on the hog," as my mother used to say, for so long, we have come to believe it is an inborn right. We have the "right" to free healthcare, the "right" to a high wage, the "right" to two cars and a big house, and the "right" to an easy life. This attitude stands in the way of doing what needs to be done to solve our debt problem.

This easy-life mentality is worsened by the political situation in the nation. Our bureaucratic systems are increasingly handcuffed by childish bickering of the worst kind. Regardless of the political party, everyone seems to be out for his own good, no matter what it means for the nation. In almost any issue you could name, if one side likes the idea, the other side hates it with a passion. The result is a system that has ground to a halt, where even the most trivial matters cannot get resolved.

What do these factors mean? They mean it is highly unlikely we will do what needs to be done to address the debt crisis we have on our hands. The barri-

TABLE 9.6 - REASONS DEBT WILL NOT BE ADDRESSED
1. Future legal spending requirements, like social security, guarantee budget increases.
2. Tax revenues in a free society are limited, meaning government income is limited.
3. The easy-life desires of Americans, and political infighting of their leaders, mean gridlock.

ers are too high and deeply imbedded to be overcome. The people of the nation, and those responsible for her governance, will not acknowledge or act on this issue until it is too late.

BUT, WE ARE TOO BIG TO FAIL!

The sum of our economic, financial, and social sins meld together into a single destructive force, the national debt. We have seen the reality of our condition in the face of Sammy Spender, an individual with the same spending and debt issues as the nation. His mirrored reflection reveals the perilous situation we have created for ourselves. Our only hope is a long-term reduction in spending and an increase in tax revenue that will allow us to slowly erode this monstrous anchor from around our necks. But forces have been set in motion that will make these adjustments nearly impossible. The heartbreaking and tragic reality is that our nation is headed for a massive economic collapse that will take the entire world down with it. How will this collapse play out, and when is it likely to occur? Our next chapter will reveal the answers.

CHAPTER DISCUSSION QUESTIONS

For a better understanding of the ideas in this chapter, work through these discussion questions in a small group:

1. Explain the difference between yearly government budget deficits and the total national debt.
2. Describe how the government borrows money.
3. Answer the following questions about Sammy Spender:
   a. How much does Sammy make a year, and how much in total does he owe?
   b. How much does Sammy spend each year?
   c. How much does Sammy spend on interest each year?
   d. At his present rate of spending and borrowing, how much will Sammy owe in 10 years?
   e. What do you expect to happen to Sammy sometime in the future from a financial perspective?
   f. How does Sammy's situation relate to the U.S. government?

4. Discuss some of the ways economists explain away the size of the U.S. debt. What are the weaknesses in these views?
5. Which of the quotes from the Founding Fathers about debt was most interesting to you? Why?
6. Describe the "simple plan for repaying the debt" discussed in the chapter.
7. Why will there be pressure on the government to increase spending in the future because of the Social Security Act?
8. Explain how there is a practical limit to how much tax income the government can collect in a free society. What does this mean for repaying the debt?
9. What did the chapter mean when it said that America had a "lack of political and social will" to address the coming debt crisis?
10. Do you think this quote from the chapter is correct? Why or why not? "It is highly unlikely we will do anything to address the debt crisis we have on our hands. The barriers are too high and deeply imbedded to be overcome. The people of the nation, and those responsible for her governance, will not acknowledge or act on this issue until it is too late."

CHAPTER TEN: THE COLLAPSE

The consequences of not dealing with our debt would be disastrous. However, given the roadblocks discussed in the previous chapter, it is difficult to see how the U.S. government will make any substantial changes to its taxing, spending, or borrowing practices in the foreseeable future. The question then becomes, What will happen to the American government and economy if we continue on this course? Let's play this out one logical step at a time.

TWO IMPORTANT QUALIFIERS

Before I outline how I believe things will transpire, let me make a couple of qualifiers. In determining what will happen to America in the future, I have received no special revelation from God. No angel has appeared to me in a dream and told me that such-and-such will happen on such-and-such a date. My conclusions have been drawn from an understanding of economics, an understanding of the Bible, and an understanding of how people think and respond in certain situations. What I am about to detail is my prayerful, educated conclusion about these matters, not divine revelation regarding them.

I also recognize I may, in the end, be wrong. Economic issues of this size and complexity are very difficult to predict, and even the most experienced and knowledgeable people can draw the wrong conclusions about them. Factors may come up I did not foresee, and people may react in ways I cannot anticipate. In fact, for the sake of the nation, and the world, I hope I am wrong.

Nevertheless, I offer these conclusions with confidence that they are in step with well understood economic and biblical principles, and that they, though frightening to consider, are not outside the boundaries of history or of basic common sense.

STEPS LEADING TO THE COLLAPSE

The first shift we will see is in the interest rates we will pay to continue to borrow money. As our debt becomes more of an issue with national and international investors, they will demand a greater return for the money they loan us. In essence, we will become a greater credit risk as our financial situation worsens, and lenders will account for that by charging us higher rates.

The move to reassess our national credit has already begun. Standard & Poor's, one of the world's foremost credit rating agencies, reduced the U.S.'s credit rating in 2011 from AAA (the highest rating possible) to AA+ due its financial condition following the Great Recession of 2008-2009, a rating that still stands today.[75] Moreover, a recent report suggests that many nations, including the U.S., could have their credit ratings reduced even further because they have "[piled] on 15-20 points of debt as a percentage of gross domestic product (GDP)—amounts that would normally take four or five years to accumulate—and locked into higher spending for the next 3-5 years."[76]

A rise in interest rates will increase the amount of the federal budget that must go to paying interest on the debt. How big an impact will this have? It depends on how much rates go up. On average, the U.S. paid 2.5% in interest on its debt between 2011 and 2020. At that rate, interest on the $27 trillion debt in 2021 will be $675 billion. This will account for about 17% of average total spending or 23% of average total revenue (recall earlier that average spending for the last 10 years has been $4 trillion and average tax revenue has been $3 trillion). In other words, when you take into account what the government will bring in through taxes *and* borrowing, the debt interest will be 17% of the total, but if you consider what it will bring in *only* through taxes (it's actual income), the percentage jumps to 23%.

If the rate jumps 1% to 3.5%, for example, the annual increase in interest will be about $270 billion, a 40% increase in interest payments. The total percentage of

75. Trading Economics, "United States – Credit Rating."
76. Marc Jones, "Exclusive: Second sovereign downgrade wave coming, major nations at risk - S&P Global." *Reuters*, 16 October 2020.

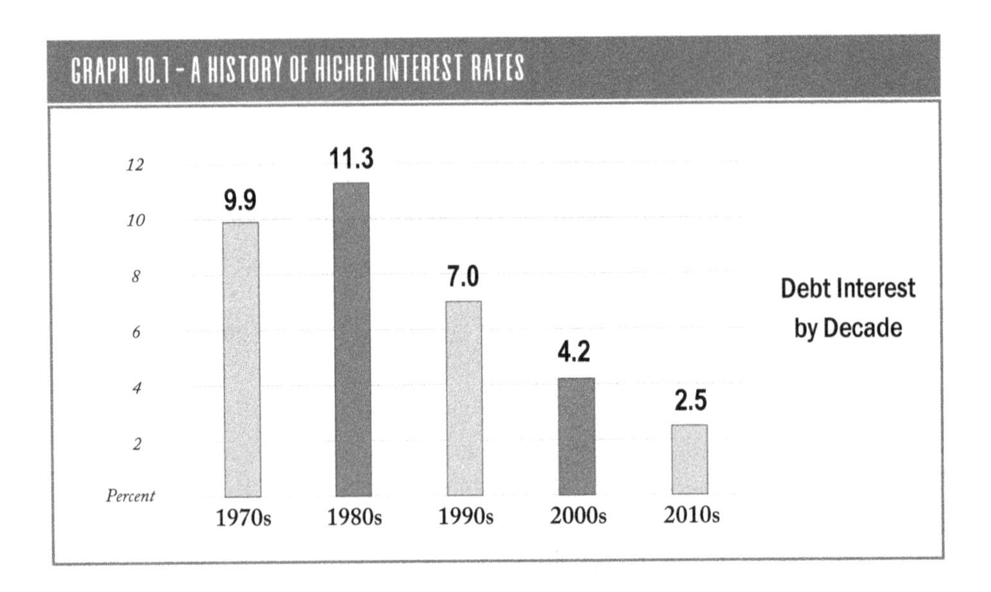

interest payments would jump to almost 24%, while the percentage of income would catapult to 32%. A 2% rise in rates to 4.5% would result in a $540 billion net interest payment increase. This would push interest payments to $1.2 trillion per year, nearly 30% of the average budget and 40% of income.

If you think this rate increase is not possible, consider that the average interest rate paid by the U.S. Treasury was nearly 10% in the 1970s, 11% in the 1980s, 7% in the 1990s, and 4% in the 2000s (see Graph 10.1).[77] Furthermore, according to the 2020 annual report of the Centre for Economics and Business Research (CEBR), a London-based think tank, interest rates are expected to rise over the next decade. "We see an economic cycle with rising interest rates in the mid-2020s," the report noted. An article on the Fox Business website suggested this likelihood would "[pose] a challenge for governments which have borrowed massively to fund their response to the COVID-19 crisis."[78] Indeed.

And what if inflation rears its ugly head, as it did in the 70s and 80s? Let's say the increased rates the U.S. must pay for higher credit risk is compounded by a boost in inflation of, say, 5%? What would happen then? We could easily end up in a situation where we are paying 9% interest on our debt, or *nearly $2.5 trillion in interest alone.* This would take up more than 60% of budget and 80% of income. (This would be like Sammy's interest payments going to $80,000 a year when he

77. The Federal Reserve Bank of St. Louis, "Federal Government Debt."

78. Fox Business, "China will overtake US as world's top economy in 2028, think tank says." 25 December 2020.

only makes $100,000 in salary.) And if interest rates reach 11%, nearly the entire income of the government—$3 trillion—would go to paying interest (see Table 10.1 for a summary of interest payments and the percent of Federal income used at each rate).

TABLE 10.1 - INTEREST RATES, PAYMENTS, & INCOME		
Interest Rate	**Interest Payment (Trillions)**	**% of Government Income Used**
2.5%	$0.675	23%
3.5%	$0.945	32%
4.5%	$1.215	40%
6.0%	$1.620	54%
7.0%	$1.890	63%
8.0%	$2.160	72%
9.0%	$2.430	81%
10.0%	$2.700	90%
11.0%	$2.970	99%

The rise in interest payments will leave the government with a severe shortfall in its budget. Yes, technically it may have the tax revenue to pay the higher interest payments, but it will still have to pay for national defense, social security, and all of the other programs it is on the hook to fund at the same time. How will it cover this shortfall? Its options are simple. It can borrow more money, raise taxes, cut spending, or some combination of the three.[79]

Borrowing will be extremely difficult because of the higher interest rates lenders will demand, and because they will be less likely to loan us all we need, depending on the size of our deficit. Even if we could borrow enough, doing so would only make the problem worse (after all, it is interest on debt that put us in this bind in the first place). Only a foolish government would attempt to solve the problem this way.

Tax increases and budget cuts are the wiser options. These kinds of measures come with one serious drawback, however—they cause the economy to slow down, and that's the last thing the government will need in a situation like this. At present, the Federal government alone makes up about 20% of the U.S. economy. Given how large a piece of the economic pie the government takes up, cutting spending by only 10% would reduce the economy by 2%. Tax increases also affect the economy. As noted earlier, federal taxes run about 18% of GDP on

79. You may be wondering why one of their options isn't to print the money to pay back the debt. It is widely understood that printing money to repay debt leads to hyperinflation. In essence, you have too much money available for the same level of production, so prices tend to rise dramatically to make up for the difference. This would destroy the economy. Therefore, it is a very unlikely option.

average. Studies suggest that a 1% increase in the tax-to-GDP level can result in a 2 to 3% drop in the economy after two years.[80]

Let's play out a spending cut/tax increase scenario. Let's say, due to higher interest rates (inflation and/or credit risk increases), the government needed to cut 10% from its spending and raise 10% more tax revenue to close its budget gap.[81] What would the result be on the economy? As already noted, a 10% spending cut would reduce the economy by 2%, but what about the tax increases? A 10% tax rate increase would add about 1.5% to the tax-to-GDP ratio. This would cut the economy an additional 3 to 4.5%.[82] All told, this spending cut/tax increase plan would cut the economy by 5 to 6.5%. Now, if that doesn't sound like much, consider that GDP only fell 4.3% during the Great Recession of 2008-2009 (see Graph 10.2).[83]

What happens when the economy shrinks like this? Tax revenue decreases because there isn't as much income for the government to tax. This results in an even greater shortfall in the government's budget. And this is when the death spiral begins. Because the government can't borrow to cover its spending needs anymore, and because its debt payments are so high, no matter what it does, things only get worse.

GRAPH 10.2 – 10% SPENDING CUT/TAX INCREASE EFFECT ON THE ECONOMY

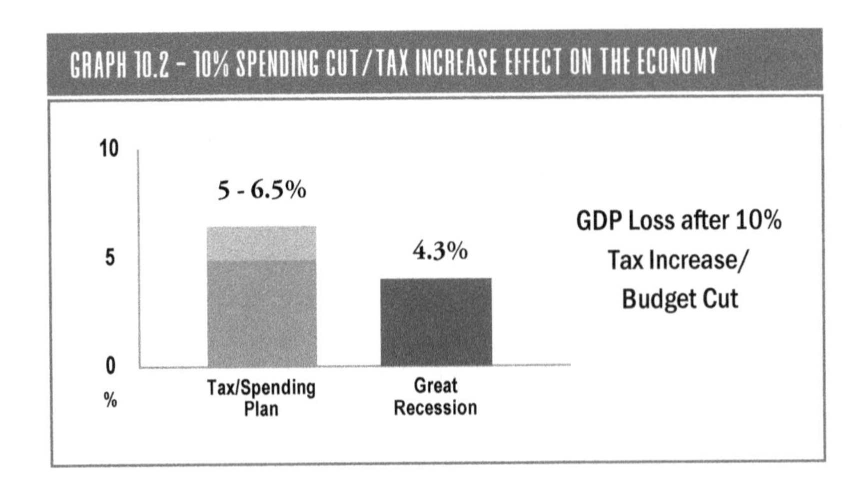

80. William McBride, "What is the evidence on taxes and growth?" *Tax Foundation*, 18 December 2012.
81. In reality, a 10% spending cut/10% tax increase action would not even get close to eliminating the budget gap if rates go as high as even 5%. But this example will help you see why we just can't cut a ton from spending and increase taxes by a huge amount overnight to solve our debt problem.
82. Remember, a 1% change in this ratio produces a 2 to 3% drop in the economy; when you multiply 1.5 times 2 and 3%, you get 3 and 4.5%.
83. Robert Rich, "The Great Recession." *Federal Reserve History*.

This is exactly what happened to Greece following the Great Recession. When times were good, the Greek government borrowed and spent too much money. When things got tight, though, government tax income dropped, lenders refused to keep lending, and the debt payments overwhelmed the government. Thomas Pugel, in his book *International Economics,* described how the Greek collapse came about:

FIGURE 10.1 - THE DEATH SPIRAL

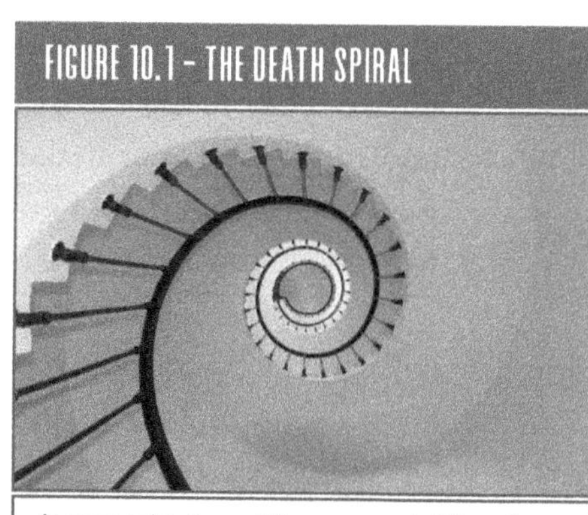

At some point, there will be no way out of the collapse brought on by our debt. We will be in a death spiral.

> After joining the euro area in 2001, Greece looked fairly successful, with annual real GDP growth that averaged about 4 percent to 2007, but its growth was based too much on fiscal deficit spending and foreign borrowing. Although the data were misreported for years by the Greek government, we now know that Greece had never met the 3 percent deficit limit. With the recession caused by the global financial and economic crisis, Greece's fiscal deficit and debt rose to high levels. By April 2010 the Greek government effectively lost access to regular borrowing—the interest rate at which the Greek government could issue new government bonds rose to prohibitively high levels. Either the Greek government had to instantly slash government spending and enact a massive tax increase or it needed an official rescue.[84]

FIGURE 10.2 - A LESSON FOR US

Greece borrowed so much its creditors would not loan it any more money, and it had to be bailed out.

The Greek government spent too much money, ran too many deficits, and borrowed too much from foreign lenders. When things were good, they were able to skate by. But when the economy

84. Thomas Pugel, *International Economics*, 17th ed., e-book version (New York: McGraw-Hill Education, 2019).

slowed down, lenders would no longer loan them money at a rate they could afford. They lost access to a big portion of their budget (the borrowed money) and could no longer make payments on their debt. In the end, they had to make massive cuts in spending and accept a rescue package from foreign governments to survive. It was a total disaster for the nation.[85]

At some point down the line, the U.S. government will meet the same fate. It will slowly be pushed further and further into a financial corner, having enormous spending obligations (like Social Security, Medicare/Medicaid, and Obamacare) pressuring from the right, shrinking income (due to tax increases and spending cuts) from the left, and massive debt payments coming up from the rear (see Figure 10.3). Every month that goes by the situation will worsen.

To slow the fall, the government may attempt to negotiate better terms with its creditors, but that will only make borrowing harder and more expensive later on. It may also implement additional rounds of spending cuts and tax hikes, but that will only increase the budget gap. It may even consider defaulting, but that will close the door on borrowing completely and may bring an end to trade with key trade partners or even war. Once the spiral begins, there is no getting out of it.

FIGURE 10.3 – FORCING THE U.S. GOVERNMENT INTO A FINANCIAL CORNER

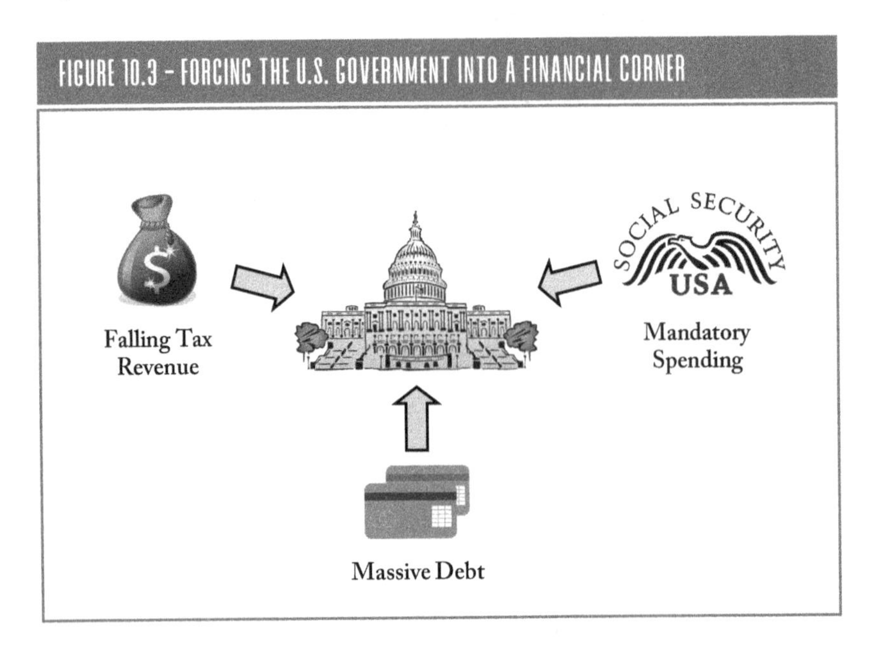

85. Also note that they misreported their debt, which led its leaders, citizens, and lenders to underestimate its debt problem. Now recall what I said earlier about the U.S. government misrepresenting the amount of interest it pays in its yearly budget. We are doing the same thing they did, in almost every way. How, then, can we expect a different outcome?

The magnitude of the financial obligations at play will not allow it. The only way it ends is in total collapse.

Perhaps you think other nations will bail us out, as they did in the case of Greece. This is highly unlikely for one simple reason: virtually all other nations depend on the U.S. for their financial and economic prosperity. When the U.S. falls in such a catastrophic fashion, the whole world will be pulled into a devastating depression with it. There simply will be no one capable of bailing us out. And remember, we are the largest economy in the world by far; it would take a massive amount of money to keep us from going under. Smaller nations, especially during a global depression, will not have the money to do it. Our foolish financial decisions will have sabotaged the entire international economy.

TABLE 10.2 - STEPS OF THE COLLAPSE
1. Interest rates increase on government debt due to credit risk, inflation, or both.
2. A severe shortfall in the government's budget.
3. Significant tax increases and budget cuts.
4. Greater budget shortfalls due to lower tax revenue and even higher rates on debt.
5. Investors stop loaning money at interest rates the government can afford.
6. The death spiral: Every move the government makes to correct the collapse worsens it.
7. A full financial and economic collapse.

HOW LONG BEFORE THE COLLAPSE?

So, when will all of this take place? There is no way to know for sure. But what history seems to suggest is that, when economic failures like this do come, they happen rapidly. Mexico had a similar economic crisis in the 1980s brought on by overborrowing and high interest rates. Dornbusch, Fischer, and Startz explain it this way: "In the 1980s Mexico went through a deep crisis. The country had borrowed too much in world markets and under the pressure of high interest rates in the early 1980s found it impossible to service [pay] its external debt. Borrowing abroad became impossible from one day to the next."[86]

Before it was all over, the result was 100% inflation, a 50% loss of currency value, a 30% decrease in wages, and a nearly 16% drop in GDP.[87] In commenting on the pace of Mexico's crisis, economist Rudi Dornbusch noted, "The crisis takes

86. Rudiger Dornbusch, Stanley Fischer, and Richard Startz, *Macroeconomics*, 13th ed., (New York: McGraw-Hill Education, 2018), 546.
87. RaboResearch – Economic Research, "The Mexican 1982 debt crisis: Economic report." 19 September 2013.

FIGURE 10.4 - OUR WORST NIGHTMARE

Inflation could cause our debt payments to skyrocket and send our government into default overnight.

a much longer time coming than you think, and then it happens much faster than you would have thought, and that's sort of exactly the Mexican story. It took forever and then it took a night."[88]

What does this mean for us? It means we will simply wake up one morning and our debt crisis will be all over the news. One day all will be fine, and the next day the dominos will fall so fast our heads will spin.

FACTORS THAT COULD HASTEN THE COLLAPSE

Although it is impossible to say when a collapse of this magnitude will take place, it is possible to identify factors that could speed up its occurrence. Here are five.

1. Inflation. As with the Mexican situation, high interest rates due to inflation could greatly hasten the collapse. Interest rates in the U.S., as reflected in debt payments, averaged around 9.5% from 1970 to 2000. Only in the last 20 years have rates been in the middle to low single digits.

Two factors in the U.S. give rise to concerns about significant inflation in the future. The first is the increasing interest among politicians in socialist policies like major wage hikes (the call for a "living" wage that is more than double the current minimum wage) and government management of significant parts of the economy (such as the healthcare system). Wage increases of this magnitude will cause prices to rise on all goods produced, and government control of major economic components guarantees mismanagement and cost overruns, both of which increase price levels.

A second concern relative to inflation is the accumulation of debt by the U.S. government. In essence, borrowing money is increasing the money supply (it brings money into the economy that was not available before). Economists have known for years that this is one of the major causes of inflation. On this point, Brigitte Granville, in her book *Remembering Inflation*, notes, "Bernholz (2003) analyzes twelve episodes of very high inflation and observes that all these episodes were caused by the financing of budget deficits through money creation. He reaches the conclusion that when the budget deficit reaches 40 percent of govern-

88. Dornbusch, Fischer, and Startz, *Macroeconomics*, 551.

ment expenditures, hyperinflation [extremely high inflation rates] will not be far away."[89] She then points out that the U.S. borrowed 40% and 42% of its budget in 2009 and 2010. In 2020, that percentage rose to 62.

FIGURE 10.5 - ANOTHER SERIOUS CONCERN

A nation with a strong economy, like China, could compete for U.S. bond investments and drive up rates.

2. *The rise of a competing economic power.* Interest rates on U.S. debt are not just affected by the government's credit rating or the level of inflation occurring at the time; they are also affected by the global market for debt. At this point in time, and for the last 75 years, in fact, the U.S. has been the safest place to put your money because it has had the strongest, most stable economy. At the end of the day, if you are an investor and are concerned about the U.S. debt situation, there aren't a whole lot of options for you—where else can you put your money that is safer? However, if, over time, as the U.S. financial condition worsens, another nation with a powerful, stable economy were to arise, the U.S. would face competition for the sale of its bonds. That competition would drive rates up. The obvious candidate for this competitor is China. The CEBR annual report referenced earlier suggested that China will overtake the U.S. as the world's largest economy by 2028. As China's economic power grows, and the U.S.'s falters, interest rates on U.S. debt may find another reason to grow.[90]

3. *Government borrowing during the early stages of the debt crisis.* As mentioned earlier, if U.S. policymakers are not wise when interest rates increase on U.S. bonds due to concerns over the U.S.'s financial condition, the collapse could jump into overdrive. Borrowing instead of cutting spending and raising taxes will cause the factors that lead to an economic spiral to build even faster. It is interesting to note that since Standard & Poor's downgraded U.S. credit in 2011, the U.S.

---

89. Brigitte Granville, *Remembering Inflation*, (Princeton, NJ: Princeton University Press, 2013).
90. This could be catastrophic for another reason as well. Right now, the U.S. issues its bonds in dollars. This means, if inflation occurs in the U.S., we still only have to pay back the same number of dollars we borrowed. However, if our condition gets so bad that we have to issue our debt in another country's currency that is more stable, say, the Chinese RMB, inflation in our nation could cause our debt payments to skyrocket. This is because we would have to pay back Chinese RMBs with dollars that are worth less on the currency market due to inflation at home. In this case, our debt levels would go up dramatically even though we had not borrowed any more money.

government has borrowed more than $12 trillion. Apparently, our political leaders haven't gotten the message.

4. Unexpected economic downturn. A fourth factor that could bring the collapse on more quickly is an unexpected disruption in the national or global economy. You need not search the history books long for examples of significant economic issues facing the nation or the world. The oil embargo of the 1970s, 9/11, the Great Recession, COVID-19—they seem to happen every decade or so at this point. Another event like these, given the U.S.'s precarious debt situation, could push us past the tipping point.

5. Investor's veil removed. A final factor that could hasten the U.S. collapse is an awakening among investors as to the true financial and economic condition of America. Today, there seems to be a veil over the minds of individuals responsible for investment decisions around the world. For some reason, they have bought into the arguments that explain away U.S. debt levels and are unwilling to see the lack of political will to correct the brewing financial crisis. They are convinced, all evidence to the contrary, that the U.S. is too big to fail. When the facts finally yank this irrational veil away, the downward economic spiral that leads to our collapse will develop quickly.

LIFE AFTER THE COLLAPSE

As suggested in the previous chapter, Americans have grown accustomed to the government always being there, checkbook in hand, when a crisis hits the nation. But what if it wasn't? What if COVID had hit and the U.S. government could not borrow a dime? To make matters worse, what if that same government during that same crisis had raised taxes and cut spending? What would life be like in that kind of scenario? Let's consider what will happen following a collapse of this magnitude.

FIGURE 10.6 - CHAOS AFTER THE COLLAPSE

The collapse of the government and economy will cause fear, chaos, and violence across America.

In the early days of the collapse, chaos, fueled by fear, will erupt across the country. Massive layoffs will occur, and unemployment applications will skyrocket. Initially, political leaders will promise to carry on as normal, providing social security, unemployment, welfare, and medical benefits as they always have. But, as the money dries up, they will not be able to keep those promises.

Benefit levels will be slashed across the board, if not completely eliminated, and people will struggle to find the food and supplies they require. Provisions on store shelves will quickly disappear, and looting will begin. Unable to buy what they need, and unprepared individually for such a crisis, people will steal what they can find from wherever they can find it. Violent clashes will result, as citizens defend their homes and businesses from mobs and thieves.

After the initial chaos, six months to a year after the collapse, new challenges will manifest themselves. Soaring inflation[91] will cause money to be increasingly useless, homelessness and tent cities will abound, food shortages will be severe, and starvation will be a real threat for many. Criminal gangs will grow in strength as cities struggle to provide a sufficient police presence to maintain law and order. In a matter of months, America will take on many of the characteristics of a Third World country.

And, while the federal government will still exist, it will be a shadow of its former self. Because of its lack of money, it will not be able to exercise the power over the states it once did. At the same time, the influence of individual states may shift, with rural, farming states becoming more prominent, while high-tech or financial-centric ones take a back seat. In order not to be pulled down by weaker states, stronger states may call for a separation from the union, potentially creating two or more nations from the former United States.

The American collapse will also shift the global power structure. It is highly unlikely that America will retain its position of dominance in the world following this event. In fact, nations ruled by dictators or communist governments may replace democratic countries as the most powerful nations on earth. Totalitarian states, at least in the short term, have a greater ability to exact money from their citizens, and, therefore, may be able to take a more influential position in world affairs.

During this time, America will be ripe for takeover by foreign powers, either peacefully or by force. If another nation or group of nations promises relief to the American people, America's political leaders, in response to the cries of citizens, may willingly pass over U.S. territory or sovereignty in exchange for assistance. Furthermore, due to the lack of funding, America's military will be greatly weakened and may not be able to repel an advance from a foreign enemy. Such a conflict may result in the first use of nuclear weapons since World War II.

---

91. It is also possible that, after the initial collapse, prices may go down for a long period of time. This is known as "deflation," and it occurred during the Great Depression when the money in the economy dried up due to hoarding and bank failures.

FIGURE 10.7 - THE WRONG CULPRIT

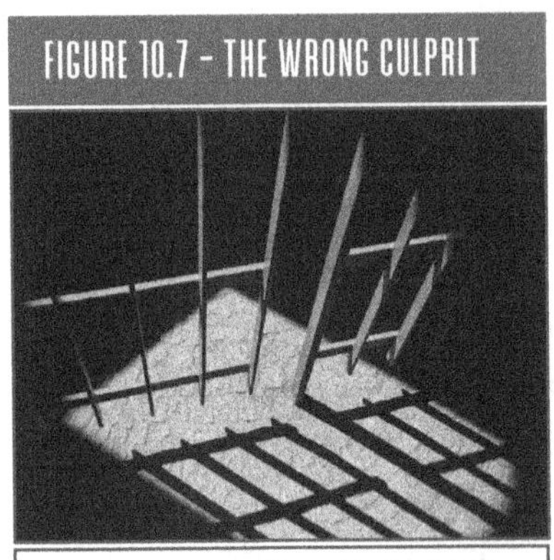

Christians could be blamed for the collapse and persecuted across the nation.

It is also possible that the Federal government may use force to maintain control of the Union, as it did in the War Between the States. Unable to collect the taxes necessary to keep the government afloat, and unwilling to let the Union fall apart, the government may use its remaining military power to force its citizens to pay the taxes demanded and the states to remain allied to the Union. A financial collapse of this magnitude has the potential to turn America into a totalitarian nation.

The collapse may also lead to a persecution of believers. Satan may be able to successfully convince the new political powers that somehow Christians are responsible for the collapse, perhaps because they would not allow the prior socialist, progressive movements in the country to take full sway. This view will not be in line with reality, of course, but that may not stop its momentum. Irrational fear, and the foolish decisions it creates, were shown clearly in the response to the COVID-19 outbreak of 2020, and both the Bible and history are full of examples of the godless twisting reality in order to harm God's people.

If the nation is able to preserve its unity, resist the influence of foreign nations, and maintain a democratic form of government, however weak it may be, life eventually will settle into a new normal. In three to five years, the people and government will adjust to the changes forced upon them. The lavish lifestyle most Americans enjoyed before the collapse, however, will not continue.

Life following the collapse will take on many of the aspects of Depression Era life from the 1930s. Several generations will live together in single homes, tent cities will become a staple of the American landscape, and jobs will be in short supply. Widespread poverty will turn many to the immoral trades to which humanity has always clung—prostitution, drugs, gambling, and human trafficking. And, though police forces will become more robust as local governments regain their footing, a strengthened criminal underworld will remain for years to come.

BUT WHERE IS GOD!

The failure of political leaders to address the massive debt of the U.S. government will eventually lead to a series of steps that will result in the destruction of the nation. Responses from lenders to U.S. debt will cause interest payments to swell, pushing the government into default. The result will be economic poverty that meets or exceeds that of the Great Depression.

But where is God in all of this? Why would he let such a thing happen to a nation with so much Bible influence in its past, and so many churches and Christians scattered across its territory? Interesting questions, indeed. Chapter 11 will reveal the brutal and terrifying truth.

CHAPTER DISCUSSION QUESTIONS

For a better understanding of the ideas in this chapter, work through these discussion questions in a small group:

1. What is the first step we will see leading to the collapse of the U.S. government? (The chapter refers to it as the "first shift.") Why is this a problem?
2. Why will a future drop in the U.S. government's credit rating worsen our debt situation?
3. What was the average interest rate paid on U.S. debt between 2011 and 2020? How did this compare to the rates it paid in the 1970s, 80s, 90s, and 2000s? What does this say about the ability of interest rates to rise in the future?
4. How much could the economy shrink if the government had to raise taxes by 10% and cut spending by 10% at the same time? How does this compare to the amount the economy dropped during the Great Recession?
5. According to the chapter, when does the "death spiral" begin for the U.S. government? Explain why this is so.
6. Summarize what happened to Greece following the Great Recession. How does this relate to what might happen to the U.S.?
7. Why is it unlikely other nations will bail us out if our government collapses?
8. Explain one of the five factors that could hasten the collapse.
9. Describe what life might be like in the early days of the collapse up until about a year following it.
10. Explain how the collapse might cause believers to be persecuted.

CHAPTER ELEVEN: THE COLLAPSE AND GOD'S JUDGMENT

I hope you have seen where we are headed, and I hope it scares you. The realities we are facing should scare you. In a very real way, this nation is headed for a crisis the likes of which it has never seen, one that could bring it to a painful and terrible end.

Now I want you to consider a deeper question: Why is all of this playing out the way it is? I don't mean the nuts-and-bolts economic reasons—spending and borrowing, interest rates and inflation. I mean from a bigger picture perspective. Is there a greater spiritual work going on here? Is God bringing this nation under judgment?

I would suggest that he is. The economic distress we are facing is a divine judgment in the making. America has offended God so severely for so long, he is preparing to strike her with a massive blow for her sins. To help you understand the details regarding this judgment, and perhaps lead you to avoid it, let's consider our dire situation through the lens of the Bible. How does God's Word explain what we see developing in our nation?

A PLACE TO START: WHO GOD IS

God is a god of judgment. Psalm 96:12-13 says, "Let the field be joyful, and all that is in it. Then all the trees of the woods will rejoice before the LORD. For He is coming, for He is coming to judge the earth. He shall judge the world with righteousness, and the peoples with His truth." The word *judge* here means "to sit as law-giver or judge in order to justify or punish." It comes from a word that can also be translated "justice." God watches the deeds of men and nations, and he acts in time to carry out justice. Of course, the Bible is clear that a full judgment of all mankind will happen at the end of the world, but it also plainly tells of God's judgments in time for man's wickedness.

FIGURE 11.1 - GOD AS JUDGE OF NATIONS

God is a god of law, justice, and judgment.

The reason God does this is because he is holy. The prophet Isaiah noted this aspect of God's character 600 years before Christ was born. In Isaiah 5:16, he wrote, "But the Lord of hosts shall be exalted in judgment, and *God who is holy* shall be hallowed in righteousness." The Creator is of a moral purity that we as fallen creatures cannot understand. Moral evil completely enrages him, because it goes against all that he is and is an attack on his very nature. Because of this, he cannot just let it pass; he must eventually act to correct it.

Fortunately for us, God is also patient beyond our understanding, and though his anger is great, he is very slow to act on it. "The Lord is merciful and gracious," the psalmist said, "slow to anger, and abounding in mercy" (Psalm 103:8). The Lord has a deep and powerful love for humanity, so he puts off his punishments as long as he can to give his creatures time to repent. However, he will not suffer our rebellion forever. Eventually his wrath passes a tipping point, and it is a frightful thing to behold. Perhaps this is why the author of the book of Hebrews said, "It is a fearful thing to fall into the hands of the living God" (Hebrews 10:31).

But who is he angry with, and why? There are two groups in America that have kindled his ire: The nation as a whole and the Christian church. Let us consider each in turn.

GOD'S ANGER WITH THE NATION

The Bible reveals a God who is in total control of the nations and disciplines them as he pleases. Job (pronounced "jobe" like "robe") said this about God and the nations, "He makes nations great, and destroys them. He enlarges nations, and guides them. He takes away the understanding of the chiefs of the people of the earth, and makes them wander in a pathless wilderness. They grope in the dark without light, and He makes them stagger like a drunken man" (Job 12:23-25). To Job's way of thinking, God does whatever he wants with the nations. The prophet

GOD'S HOLINESS AND JUDGMENT

God never sins, and, in fact, cannot sin. He is perfectly pure from a moral standpoint. Because of this, he hates sin and must eventually punish it.

Daniel agreed. "All the inhabitants of the earth are reputed as nothing," he declared. "[God] does according to His will in the army of heaven and among the inhabitants of the earth. No one can restrain His hand or say to Him, 'What have You done?'" (Daniel 4:35).

Throughout biblical history, God intervened in the affairs of nations, punishing some and rewarding others. Among the nations he disciplined were Egypt, Assyria, Babylon, the Ammonites, the Philistines, the Edomites, Syria, Samaria, Elam, Kedar, Hazor, and, of course, Israel. He also accomplished great deliverances for the Philistines, Syrians, Ethiopians, and Israelites.

Though the Bible's pages are now written, there is no reason to believe God has left the nations to themselves. The Bible's truths about God's judgment are as applicable today as they were during the time of Moses, David, Jesus, and Paul. We are about to learn this firsthand.

America's history with the God of the Bible

Why, then, is God angry with America? Understanding the answer to that question begins by understanding America's history with the God of the Bible. While the mainstream American media, entertainment, and educational systems would have you believe this nation had only a passing relationship with the Christian God in its early years, the truth is quite the opposite.

One of the most thorough and powerful reviews of America's history with God is found in a Supreme Court decision from 1892, the case of the *Church of the Holy Trinity vs. the United States*. This case centered around a law passed by congress that limited the foreign workers that could enter the U.S. for employment. Apparently, the Church of the Holy Trinity wanted to hire a minister from England, a Mr. E. Walpole Warren, but Mr. Warren was not allowed to enter the country because of this law. The case made it to the Supreme Court in January of 1892. The court handed down its decision at the end of February of that same year.

FIGURE 11.2 - A LONG RELATIONSHIP

America has always had a respectful and loving relationship with the God of the Bible and Jesus Christ.

The justices decided in favor of the church and struck down the interpretation of the circuit court that the law could be applied to clergy. In the decision, Justice David J. Brewer described the legal justification for the decision, and then made what would be to the modern American ear a shocking statement: "But beyond all these matters no purpose of action against religion can be imputed to any legislation, state or national, because this is a religious people." In other words, the court was saying, even if the legal tests had not been met, there is no way a law passed by the U.S. Congress could be interpreted against a Christian church because America was inherently and thoroughly a Christian nation.

FIGURE 11.3 - JUSTICE OF HISTORY

Justice David J. Brewer detailed America's history with the Christian God in an 1892 case.

He then goes on to recite the clear and powerful Christian influence on the founding of the nation. This influence can be seen in the commission of Christopher Columbus by Ferdinand and Isabella, the first colonial grant given to Sir Walter Raleigh by Queen Elizabeth, and the first charter of Virginia bestowed by King James I. Justice Brewer then reviews a similar Christian influence in the establishment of the first colonies, the constitutions of the first 44 states (all of them at that time), the Declaration of Independence, and the U.S. Constitution.

The whole of his argument can be summarized in the following quotation found toward the end of his remarks (the text of the entire decision can be found in Appendix A):

> There is no dissonance in these declarations. There is a universal language pervading them all, having one meaning; they affirm and reaffirm that this is a religious nation. ... While because of a general recognition of this truth the question has seldom been presented to the courts, yet we find that in *Updegraph v. The Commonwealth*...it was decided that "Christianity, general Christianity, is, and always has been, a part of the common law of Pennsylvania...."
>
> And in *The People v. Ruggles,*...Chancellor Kent, the great commentator on American law, speaking as Chief Justice of the Supreme Court of New York, said: "The people of this State, in common with the people of this country, profess the general doctrines of Christianity, as the rule of their faith and practice; and to scandalize the author of these doctrines is not only, in a religious point of view, extremely impious, but, even in respect to the obligations due to society, is a gross violation of decency and good order." ...

> If we pass beyond these matters to a view of American life as expressed by its laws, its business, its customs and its society, we find everywhere a clear recognition of the same truth. Among other matters note the following: The form of oath universally prevailing, concluding with an appeal to the Almighty; the custom of opening sessions of all deliberative bodies and most conventions with prayer; ...the laws respecting the observance of the Sabbath, with the general cessation of all secular business, and the closing of courts, legislatures, and other similar public assemblies on that day; the churches and church organizations which abound in every city, town and hamlet; the multitude of charitable organizations existing everywhere under Christian auspices; the gigantic missionary associations, with general support, and aiming to establish Christian missions in every quarter of the globe.
>
> These, and many other matters which might be noticed, add a volume of unofficial declarations to the mass of organic utterances that *this is a Christian nation*. [emphasis added]

Whether we like it or not, America, from the first day Columbus set foot on its shores, through the establishment of its colonies, states, and union, to the middle part of the 20th century, was a nation built upon respect for and admiration of the Bible, its God, and its central figure, Jesus Christ. As we reflect on the blessings bestowed upon this country—the rich agricultural land she enjoys; the technological advancement she has known; the prosperous economy she has developed; the great wars she has won; the global stability she has brought—can we deny the favor of this same God, a clear token of his pleasure with us as a people?

FIGURE 11.4 - THE MOCKED GOD

Atheists and the wicked are very bold in their hatred of God.

modern America's sins

Now, against that backdrop, let us consider our nation today. Not only has this God been pushed out of public life, his very existence has been challenged openly on billboards and road signs (Figure 11.4), in classrooms and media broadcasts, from sea to shining sea. When his existence is not denied, he is mocked, ridiculed, and demonized by an increasingly godless society. As evidence of this, consider how the name of Jesus Christ has become a curse word akin to the vilest terms in the language, an insult the Chief Justice of New York called "a gross violation of decency and good order."

FIGURE 11.5 - WICKED AND UNASHAMED

American's flaunt their perverse sexual ideas, like homosexual marriage, transgenderism, and polygamy.

Moreover, the Creator's definition of the family has been erased from the public mind, replaced by couples living together as "partners," women working while men stay at home, and homosexuals exchanging marriage vows with full approval of law. Sexual intimacy, designed to be enjoyed in the one man/one woman marriage relationship, has become a perverse web of debauchery. Strip clubs, escort services, prostitution, pornography, men with men, women with women, group sex, swingers, "friends with benefits"—no sexual practice is off limits to the modern American.

And, while the Bible values the life of every person, America murders 800,000 children in their mothers' wombs every year, over 60 million since 1973. If that was not bad enough, women in America are beaten by their boyfriends, children are molested by their fathers, parishioners are shot in the pews, and teenagers are murdered in the schools. With so much death surrounding us, we would do well to remember the words of King David from Psalm 9:17, "The wicked shall be turned into [the grave], and all the nations that forget God."

But this is not the end of our sins. While modern Americans denounce Christianity, they openly embrace every idolatrous substitute the world has to offer. We applaud the deadly religion of Islam, promote the inhumane caste system of Hinduism, and rejoice in the meaningless principles of the New Age. We build temples to science, worship evolution, and stand awestruck at the glory of man's reason. And when these idols do not have our full attention, we bow before arrogant sports figures, accept the words of uneducated actors, follow the lifestyles of corrupt singers, and honor the decisions of villainous politicians.

All the while, the American people become more and more lawless every day. The new rule in America? *Whatever you can get by with.* We lie to our spouses, defraud our government, cheat our employers, manipulate our friends, exploit our children. And when the opportunity presents itself, we riot over offenses that never occurred, burn down businesses that others have built, and cry out for jus-

FIGURE 11.6 - DO YOU REMEMBER 9/11?

God has been punishing America since 9/11.

tice while we ourselves carry out injustice upon others. We are a nation of takers, not givers; extortioners, not helpers; militants, not peacemakers.

And you ask why God is angry?

trying to get our attention

In fact, he has been showing his anger for more than 20 years, but we have paid him no attention. Who recalls the shock of 9/11? Who remembers the airplanes full of people flying into the World Trade Center towers? Can you still visualize the flailing bodies as they fell from its smoke-filled windows? And do you recall who we turned to in the first days after that tragedy? What was the motto of all America in the first few days following these attacks? President George W. Bush quoted it in an address to congress a week after that fateful day. "All of America was touched on the evening of the tragedy," he said, "to see Republicans and Democrats joined together on the steps of this Capitol singing '*God Bless America*.'" But how quickly that faith faded, and our new motto became, "United We Stand," reflecting a trust and dependence on ourselves instead of God.

The Great Recession of 2008-2009 was another shocking blow by the Almighty. Almost overnight, our entire banking system nearly crumbled. It was the worst economic downturn since the Great Depression. Let that sink in for a moment—*the worst economic downturn since the Great Depression*. Yet, have we returned to God?

Over the last two decades, the Creator has also hammered us with nature. Reflect on the fires consuming California. One report said over four million acres had been burned, with the loss of 31 lives and over 8,500 structures. Scott McLean, a spokesman for Cal Fire, said, "The four million mark is unfathomable. It boggles the mind, and it takes your breath away."[92] Yes, it takes our breath away, but apparently it does not drive us to our knees.

And fires are just the beginning.

In 2019, the National Centers for Environmental Information compiled a list of 115 natural disasters that had impacted the U.S. in the previous 10 years, each

92. Wynne Davis, "California Wildfires Have Burned 4 Million Acres And The Season Isn't Over Yet." *NPR*, 4 October 2020.

of which cost more than $1 billion.[93] Tornadoes, floods, hurricanes, droughts—you name it. Monikers like "Superstorm Sandy," "Bomb Cyclone," and "Snowpocalypse" represent significant weather-related events that have struck the nation. And who could forget Hurricane Katrina in 2005, a massive storm that devasted New Orleans and the surrounding area (Figure 11.7). While on the subject of hurricanes, in 2020 alone "twelve named storms, including a record-tying six hurricanes, have made landfall in the United States."[94]

FIGURE 11.7 - NATURAL DISASTERS

The damage left by Hurricane Katrina is but one of many natural disasters God has sent to America.

And then the strongest blow yet—the global pandemic of 2020. The entire nation shut down for weeks as fear, confusion, and death spread across the land. Remember how you felt in those first few months? Do not let those emotions fade. What a blow to a nation—the whole world, even—that God struck at that time.

we are seeing God's work

This is exactly what God does to a nation that has rebelled against him. Drought was used in Jeremiah 3:2-3: "And you have polluted the land with your harlotries and your wickedness. Therefore the showers have been withheld, and there has been no latter rain." Fire was the instrument in Numbers 11:1, "Now when the people complained, it displeased the Lord;...and His anger was aroused. So the fire of the Lord burned among them, and consumed some in the outskirts of the camp." Plagues were the order of the day for the Egyptians, including a disease that affected the skin with boils and ulcers: "Then they took ashes from the furnace and stood before Pharaoh, and Moses scattered them toward heaven. And they caused boils that break out in sores on man and beast" (Exodus 9:10). And, to bring down the mighty, arrogant nation of Moab, the Lord took its wealth, as Jeremiah reveals:

> Flee, save your lives! And be like the juniper in the wilderness. For because you have trusted in your works and your treasures, you also shall be taken. Therefore My heart shall

---

93. Katelyn Newman, "Unforgettable U.S. Natural Disasters of the 2010s." *US News*, 27 December 2019.
94. Jason Samenow, Andrew Freedman, and Matthew Cappucci, "2020 Atlantic hurricane season breaks all-time record while leaving Gulf Coast battered." *Washington Post*, 10 November 2020.

> wail like flutes for Moab, and like flutes My heart shall wail for the men of Kir Heres. Therefore *the riches they have acquired have perished.* **Jeremiah 48:6-7, 31, 36**

The judgment of Moab will be our ultimate downfall. The Mighty God will strip us of our riches in a dramatic and horrible fashion. The floods, fires, droughts, and storms we have seen thus far are but gentle previews of what is to come.

GOD'S ANGER WITH HIS PEOPLE

But God's anger is not just directed at the nation as a whole. It is specifically pointed at his own people, the Christian church. After all, the godless are only doing what they were always going to do, turn from God and fight against him with all their might. But his people are to know better, to do better, to be better. And yet, they have failed him in nearly every way.

In fact, it could be said that the fall of the nation is a direct result of the fall of the churches. The great influence they had in the first 150 years of American history has been completely lost. The tremendous spiritual starting place granted this nation, one that has never been seen outside of Israel itself, has been criminally mismanaged and destroyed.

The central sin of Christian people in America is a lack of faithfulness. Slowly, our allegiance and obedience to Christ has slipped, and our service has transitioned to a new master—ourselves. We have dismissed *his will* for what we are to do and to be for what *we want* to do and to be.

Consider the life and work of our Lord Jesus Christ. What did he do when he was on the earth? He aggressively took his gospel to the lost, he sacrificially met the needs of the poor, and he courageously stood for the truth of God before the lawless. But what about us?

Meditate on our evangelism efforts. How many churches are really seeking the lost as Christ did? How many make his gospel their primary work and mission? I have been in the church for more than 20 years, have pastored three churches myself, and have spoken in many congregations across the country. What I can tell you without a moment's hesitation is that the Christian church in America, for all practical purposes, has completely abandoned the work of seeking and saving the lost. Our churches are filled with white hair and empty pews precisely because of our stubborn refusal to do what our Master told us to do—go and make disciples.

This can be seen statistically, as well. John Ewart, a Southern Baptist consultant, seminary professor, and former pastor, noted the startling percentages of average believers who are not willing to share their faith. Pay close attention to his words:

> I could show you easily a dozen studies that say that well over 90% of our active church members have no intention whatsoever of ever sharing their faith. It's not just the fact that they don't share their faith; they have no intention of sharing their faith. Some of those go from 90% to 96%, and I argue from one that it is as high as 98%. Whatever the number is, here's my point: I do not believe that the active evangelistic labor force is large enough to impact this culture.[95]

And what about our work for the poor? In modern America today, where do the hungry, the sick, and the oppressed go for help? The churches? Not a chance. The U.S. government—that's where. In the grand scheme of things, if the Protestant church in America ceased every program it had for the needy today, there hardly would be a ripple in the big pond of provision for the destitute. What an everlasting shame that the ones who are supposed to be the best friends of the down-and-out are hardly a blip on the radar of their daily lives.

We find perhaps our greatest failure when it comes to standing for truth. One at a time, our culture has rewritten the foundational moral principles of this nation, all of which came from the Bible, right under our noses. And what have we done about it? We used to put up a fight, take a stand on the Bible's teaching, defend it before the masses of haters, and serve as an important barrier to the ruining of the nation. But now, after generations of pressure and misinformation, we have given in to their arguments and compromised God's Word at every turn.

Our people have found ways to say the Bible approves of homosexuality, that intercourse before marriage is not that bad, that God "loves us too much" to keep us from marrying whatever sex we want. We have given in to evolution and affirmed that God used millions of years of death and suffering to create life. We even write books to talk about how "good" death is

GOD'S ANGER AT CHRISTIANS

God is angry with American Christians because they refuse to obey him, believe the lies of the world, and have become the friends and supporters of the godless.

95. John Ewart. Level 1 Training DVD Course. Louisville, KY, *Society for Church Consulting*, 2008.

FIGURE 11.8 – FALSE PROPHETS

Christians like Hugh Ross affirm evolution and write books about how death is good.

(see Figure 11.8). And our pulpits, instead of taking the heat for the truth, steer clear of any and all controversies so as not to offend the valued "seeker."

Worse than this, in recent years, we not only agree with the godless; we publicly take their side. I have a Christian friend here in South Carolina who claims to know the Lord. He talks about God, reads the Bible, and prays. Yet he openly supports abortion because "they're going to do it anyway." I have another Christian friend in North Carolina. He attends an evangelical church regularly and even plays in a Christian band. He is probably 60 years old. I got this message from him recently on Facebook about homosexual marriage: "If gays and lesbians want to be married, sin or not, why should our laws outlaw it?" Countless thousands of God's people even cast their votes for Democrat party candidates in local, state, and national elections.[96] Far from the enemies of the godless, we have become their greatest allies.

This is one of the reasons God destroyed his people during the Old Testament period. Jeremiah 23:14 reads, "Also I have seen a horrible thing in the prophets of Jerusalem: They commit adultery and walk in lies; they also *strengthen the hands of evildoers*, so that no one turns back from his wickedness. All of them are like Sodom to Me, and her inhabitants like Gomorrah." The prophets, who were supposed to stand with the Lord on all matters of truth, had become so corrupt that they were "strengthening the hands of evildoers." In other words, they were telling them that what they were doing was good and that God approved of it. What did these deceived believers receive for this? They were destroyed like Sodom and Gomorrah.

And while we are refusing to do Christ's work and publicly taking the side of his enemies, what is going on in most of our churches? Infighting, selfishness, greed, adultery, divorce, and hard-heartedness. Some of the most hurtful, dysfunctional, and wicked places I have ever seen are Christian churches.

---

96. This is not to say that the Republican party, or any other party, is righteous and in God's favor. In terms of the Bible, no political party is righteous. They all stand for things God hates. However, the Democrat party outwardly professes and aggressively pursues godless ideas and policies of all sorts, including abortion, homosexual marriage, and the persecution of Christ's people. For a Christian to vote for such a group is contrary to all biblical principles and is a sin against God.

As with the national judgments mentioned earlier, the Bible clearly reveals that God judges his people when they sin. In the first few chapters of Revelation, the Lord promises the following punishments for disobedient churches and believers: to remove them from the kingdom (2:5); to fight against them with the sword of his mouth (2:12); to bring sickness, catastrophe, and death to their members (2:22-23); to spit an entire church out of his mouth (3:16); and to do all of this suddenly and without notice (3:3).

In other places, we see the severe chastisement of God against his people. In Acts 5:1-11, the Spirit of Jesus killed two Christians for lying to him, and, according to Paul, several members of the Corinthian church had been made very sick or put to death for sinning during the Lord's supper (1 Corinthians 11:30). And need I mention the 603,550 Israelites who died in the wilderness for refusing to enter the land of Canaan? Here is what God himself said to his disobedient people at that time, "The carcasses of you who have complained against Me shall fall in this wilderness, all of you who were numbered, according to your entire number, from twenty years old and above" (Numbers 14:29). God is a god of grace and patience to the humble and obedient, but he is a god of wrath and vengeance against the godless and wicked, whether they claim his name or not.

WHAT DO WE CONCLUDE FROM ALL OF THIS?

When we look at our situation through a biblical lens, and when we consider the rich Christian heritage of this nation and the degree to which believer and unbeliever alike have disobeyed and offended the God of the Bible, there is absolutely no doubt that the catastrophe that awaits us is of God. His anger has reached a rolling boil, and he has already prepared the instrument of his justice. If the nation and its churches continue on the course they have been on for the last 75 years, a great humbling, one like nothing this land has ever seen, is coming to every corner of this land. Our prideful arrogance will be ground to powder, and every man and woman, every institution and religion, every race and social class will be smashed into dust.

FIGURE 11.9 - ONLY ONE ANSWER

The only hope to stop God's hand of judgment upon America is widespread and complete repentance.

Our one and only hope is repentance. As I have tried to show in this book, the economic, political, and social realities in our nation virtually guarantee we will not make the changes necessary to delay or prevent our collapse. The single remaining opportunity to prevent judgment is to turn to the God of the Bible in true, deep, and complete repentance.

If you are not a Christian and are reading this book, get down on your knees right now and beg God for the forgiveness of your sins against him. As a man or woman outside of God's grace, you not only face the judgment coming to America; you face the Final Judgment of the world at the end of time. And the punishment received after that judgment will make what happens to America in the coming years seem trivial. Christ is your only hope for forgiveness and the only chance you have to receive eternal life. He died to pay your sin debt so you could be free. Turn to him, repent, put your faith in him, and use the rest of your life to serve him. (See Appendix C for more on how to be saved.)

If you are a professing Christian, return to Christ through repentance now. Acknowledge the ways you have failed to serve him faithfully, believe his Word, and stand for his truth. Ask for forgiveness for siding with his enemies and allowing yourself to be deceived into thinking the lies of this world are the truth of his Word. If you do not do this—if you continue in your hardheaded belief that you are right—you, regardless of your empty claims of faith, will be punished with the rest of unbelieving America, and on the great Day of Judgment at the end of time, you will be condemned to hell.

What I say to one, I say to all, repent.

BUT WE'LL ALL BE DESTROYED!

God is holy and just, and he will not endure man's wickedness forever. America is well acquainted with the God of the Bible, and her sins against him are about to be punished in a severe way. The only hope for her now is repentance.

But how will we survive such an awful event? Will this be the end of God's people in America, or is there hope for us yet? And is there anything we can do to prepare ourselves before the collapse takes place? Read on. One chapter of answers remaining.

CHAPTER DISCUSSION QUESTIONS

For a better understanding of the ideas in this chapter, work through these discussion questions in a small group:

1. Is God a god of justice and judgment? Support your answer with verses from the Bible.
2. How is Psalm 103:8 a source of hope for the future of our nation?
3. Does God still rule the nations? Support your answer with verses from the Bible.
4. What were you most surprised to learn about America's history with the God of the Bible? Had you ever heard any of these things at school?
5. What are some of the sins of America that God may be judging through the collapse?
6. How might God have been trying to get our attention as a nation over the last 20 years? What judgments has he brought on us that we have not recognized as such?
7. Why is God angry with his own people in the U.S.?
8. Do you think this statement from the chapter is true or false? Why? "It could be said that the fall of the nation is a direct result of the fall of the churches. The great influence they had in the first 150 years of American history has been completely lost. The tremendous spiritual starting place granted this nation, one that has never been seen outside of Israel itself, has been criminally mismanaged and destroyed."
9. How have American Christians been publicly taking the side of the wicked? Have you noticed this? Share an example if you have.
10. What is the only act that can prevent God's judgment of our nation? What can you do to keep judgment from coming?

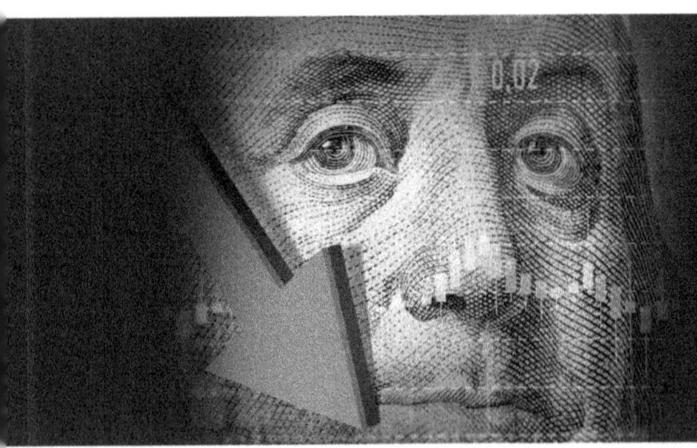

CHAPTER TWELVE: PREPARING FOR & PROSPERING THROUGH THE COLLAPSE

The prospect of so great a judgment can create overwhelming fear, apprehension, and anxiety. God has placed in each of us a desire to prosper and do well, and when we anticipate a period of terrible trial ahead, it can fill us with hopelessness and dread. This response, however, is not necessary for the believer. In fact, the very opposite is possible for the follower of Christ.

Times like the ones we are faced with will certainly bring difficulties to believers, but in the middle of those difficulties, an amazing time of prosperity can be found. It is one of the great paradoxes of the Bible that God's people would be promised, at the same time, so much affliction and so much joy. Consider the amazing promises of the Bible to God's faithful servants during times of great distress:

> The LORD knows the days of the upright, and their inheritance shall be forever. They shall not be ashamed in the evil time, and *in the days of famine they shall be satisfied.* **Psalm 37:18-19**

> Behold, the eye of the LORD is on those who fear Him, on those who hope in His mercy, to deliver their soul from death, and to *keep them alive in famine.* **Psalm 33:18-19**

> The LORD *will not allow the righteous soul to famish,* but He casts away the desire of the wicked. **Proverbs 10:3**

> He shall deliver you in six troubles, yes, in seven no evil shall touch you. *In famine* He shall redeem you from death, and *in war* from the power of the sword. And you shall not be afraid *of destruction* when it comes. You shall *laugh at destruction and famine.* **Job 5:19-22**

There is no doubt that, should God complete the judgment he is preparing, dark days will come. As has already been pointed out, famines, violence, and war are reasonable outcomes of such a collapse. But these verses promise, even in the middle of such catastrophe, that God will look out for his people.

Famine—one of the worst situations that could be imagined—will not destroy Christ's own. In fact, as Job indicated, the righteous will laugh at such circumstances, because their helper is the Creator himself. After all, what famine is too great for God? What army or violent mob can harm that which he protects? Interestingly enough, the word translated "satisfied" at the end of Psalm 37:19 can be translated as "filled, have one's fill, or have in excess." An absolutely unbelievable promise of hope for God's people, that, in the time of famine, they would have excess.

What does this mean for us? It means we can enter into this period of judgment with the hope that God will care for us during it. This does not mean we will not suffer, or that we will not endure losses that bring us sadness. What it does means is that, ultimately, God will provide a way for us to prosper, and will only allow us to suffer that which he will turn for our good. The time we are headed for can be a great time of progress and joy in his kingdom. Lay hold of these promises and allow them to guide your thoughts and actions in the dark days ahead.

EDUCATE OTHERS AND STEM THE TIDE

Before discussing how we should prepare, let me share a word about what you can do to educate others about the coming collapse. First of all, realize that God may have awakened you to our situation so that you can open the eyes of others. It is important that we try to inform those around us of the things we have learned about America's financial and economic situation. Share this book with your fellow Christians and your pastor. If you are a teacher in the church, use it as a basis for a Sunday school, Bible study, or sermon series. (For teaching resources related to this book, visit www.wesmoorenow.com.) Hand out this book at Christian gatherings, conventions, and events. Do your part to wake up the church. Perhaps more will repent, and God will hold back his judgment.

FIGURE 12.1 - LET OTHERS KNOW

Share the message of this book with other Christians.

Also, it is critical that you make the national debt the most important issue in your political activity. Research candidates' positions on the debt and vote for those who will try to limit or even reduce it. If you are in any kind of elected office, share the truths about our debt and the likely consequences of continuing on our present path. You never know how God might use you to bring light and deliverance to others in the days to come.

PREPARING FOR THE COLLAPSE

Part of the way God will provide for us during and after the collapse is by instructing us on how to prepare for it in advance. As in the days of Joseph when God foretold of the years of famine that were to come, prosperity during difficulty begins by getting ready for it before it occurs. So, what should we do now?

Before I get into recommendations for preparations, let me say that I know some of these are going to sound like they came out of an apocalyptic movie. I am not trying to be extreme here. I am only trying to think through what would be necessary if things go as badly as we reasonably suspect they may. Those of us who know the Bible, know that mankind is not good at its heart. If the economy falls apart, and the government loses its grip on the population, we could very well see some things we never thought possible in our country. Read the following recommendations with that in mind.

truth-centered prayers

Our preparation should begin with prayer. There is so much to pray for at this time. The first thing we should ask God for is mercy for our nation. He has shown his willingness in the past to stay judgment when his people plead with him for it (Isaiah 37). We should beg him to glorify his mercy instead of his judgment and to remember the suffering of his people as the day of justice approaches. We can also reason with him regarding the harm that will be done to children, the disabled, and the elderly as a reason to forgo this punishment. He has shown a willingness to withhold judgment for this reason in the past (Jonah 4:11).

These prayers for mercy must extend to the granting of repentance to the church and the lost within our borders. Just this morning, I was reading the book of Jeremiah and found some amazing words of hope for a nation about to endure God's judgment. Listen to this promise from God:

> In the beginning of the reign of Jehoiakim the son of Josiah, king of Judah, this word came from the Lord, saying, "Thus says the Lord: 'Stand in the court of the Lord's house, and speak to all the cities of Judah, which come to worship in the Lord's house, all the words that I command you to speak to them. Do not diminish a word. Perhaps everyone will listen and turn from his evil way, *that I may relent concerning the calamity which I purpose to bring on them because of the evil of their doings.*'" **Jeremiah 26:1-3**

FIGURE 12.2 - A FOUNDATIONAL ACT

Preparation for the collapse begins with serious prayer.

For years, God had promised Judah a devastating judgment for its sins. In these verses, yet again, he is sending his prophet to call them to repentance. Why? *Because he did not want to do what he had planned to do*. Oh, what a merciful and kind being God is.

In the end, we know that only God can grant repentance (2 Timothy 2:25), so we must beg him to do so that we might be spared this destruction. If he is willing to grant that repentance, we must also ask for wisdom for our leaders to make the decisions necessary to steer us clear of this crisis and the willingness of our citizens to support those decisions.

Our prayers should then shift to wisdom and blessing in preparing plans and supplies for the collapse. This event will undoubtedly create circumstances and problems no human could anticipate. We need God's infinite knowledge to assist us in getting ready.

Finally, we should pray that, if God moves forward with this judgment, that the excellence of his wrath, justice, and holiness would be made known to our people and the people of the world. God's purpose in all that he does, even judgment, is to reveal his perfections to all of creation (Romans 9:17). If he, in his infinite wisdom and goodness, determines that this event

TABLE 12.1 - WHAT TO PRAY ABOUT
1. That God would show mercy to our nation, bring repentance, and prevent the collapse.
2. That God would grant us incredible wisdom so that we might prepare for the collapse.
3. That, if God wills the collapse, his perfect wrath, justice, and holiness be made known.

FIGURE 12.3 - HELP WITH DEBT

Dave Ramsey's financial ministry can provide detailed guidance for getting out of debt.

must take place, we must honor his will and pray that his purposes, though difficult for us to grasp, are accomplished.

wise preparations

As our prayers continue, we must take up the task of preparing what we will need during and after the collapse. These preparations fall under two headings, individual and church wide. Let's review these now.

individual preparations

Individually, you must begin by getting yourself out of debt. You need to ensure that, to whatever degree possible, the things you need, you own outright. It is especially important to pay off credit card debt, vehicles, and homes. Depending on how much time the Lord gives us, you may not be able to pay off everything, but you must do all you can.

Americans have developed so many bad habits, many do not know how to become debt free. If you fall into this category, I highly recommend Dave Ramsey's financial teaching ministry (see Figure 12.3). Search for him on the internet and put his principles into practice.

Secondly, you must build up cash in something other than a bank. High inflation is likely in the initial stages of the collapse, but if you have no cash at all, you will have very little means to purchase the things you need. A financial collapse of this magnitude may cause banks to go under, and with the Federal government out of money, it may not be able to insure your deposits. It is therefore wise to begin to store cash somewhere other than a bank that is safe and that you can access in a time of crisis. I do not have a recommended amount. I only suggest you put away all you can.

You will also need to store food and other provisions. Supply lines will likely break down during and after the collapse. You will need food, soap, medicine, candles, and other supplies to keep your family afloat during this time. Having a working generator and supply of fuel would be wise as well. I recommend a two-year supply of everything. As with your money, store these supplies in a safe place that you can easily access in an emergency. Also, be sure to consider methods to increase the shelf life of the food you store. You can find this information on the internet.

Regarding food, you should give some thought to how you could grow food after the collapse. Most of us know very little about gardening. However, if we can develop the skill between now and then, it would be helpful for our survival afterwards. Read up on growing vegetables, and plant a small garden in your backyard. You may want to buy a few gardening books that you can store with your emergency supplies. And it may be wise to store some seeds of common vegetables so you will have something to plant. Buying these kinds of items from stores after the collapse may be difficult.

FIGURE 12.4 - SET ASIDE PROVISIONS

Prepare two years of food and other provisions.

If you have the means, you may consider buying property somewhere out of the way of big cities. You can use this property to grow food if you need to, and, if you have some kind of home on it, you can use it as an emergency location for your family. Not everyone can do this, but if you can, it could be a lifesaver.

Wherever your family will be during the crisis, you need to have weapons ready to defend yourself. You're not preparing to fight a war, just trying to keep your family and supplies safe in case of riots and looting. Be sure to get some training on using those weapons, and always store them in a safe place you can get to if things get out of hand.

Finally, you should develop an emergency plan for your family and make sure your adult family members know about it. Lay out what your provisions are, where they are, and how to get to them. Specify what you are going to do if something happens. Are you going to stay put and use your provisions, or are you going to try to get to someone else's house or to an emergency location? You need an action plan so you can protect yourself and those you love if something happens.

preparations for the churches

Churches need to mirror most of the preparations of individuals. They should get themselves out of debt, build up cash reserves, store up food and supplies, and develop an emergency plan. They may also buy or develop land that could be used to grow or

TABLE 12.2 - INDIVIDUAL PREPARATION LIST
1. Get out of debt; own everything you need.
2. Build up a cash reserve outside of a bank.
3. Store two years worth of food and supplies.
4. Buy property outside of cities to use as an emergency location for you and your family.
5. Develop an emergency plan for your family.

FIGURE 12.5 - AN EMERGENCY PLAN

Create and communicate an emergency plan.

store food. Special precautions must be taken to ensure these areas are secure.

In addition, like-minded churches should ban together to support one another. Church leaders should meet to discuss how they can coordinate activities, share services and skills, and provide protection for the resources they have amassed. Plans of action should also be drawn up. Exactly what will the churches do if the economy falls apart? What are the exact steps they would take to provide for their members? Thorough planning now will pay great dividends later.

PROSPERING AFTER THE COLLAPSE

This collapse, should God see fit to send it, will not be the end of God's people. Flip through the pages of the Bible and you will find countless catastrophes the people of God encountered and survived. The believer knows that Paul's words are actually true: "Who shall separate us from the love of Christ? Shall tribulation, or distress, or persecution, or famine, or nakedness, or peril, or sword? … Yet in all these things we are more than conquerors through Him who loved us" (Romans 8:35, 37). Yes, we are more than conquerors through Jesus Christ, and we will be more than conquerors through this as well. As I said in the opening paragraphs of this chapter, God will not only see us through this dark time, he will also *see that we prosper* through it.

Consider some of the good that will come through this catastrophe, good that otherwise could not be possible. First of all, we will be freed from our materialism. It is hard to see now, but our prosperity has become a noose around our necks. We work so hard for more and more things, and in the process lose the true meaning of life and the ability to enjoy it.

FIGURE 12.6 - AN UNKNOWN BLESSING

The collapse can free us from the power of materialism.

Life will be simpler after the collapse. I'm not sure we can fully understand how stressful and busy our lives are today, or the toll that that kind of lifestyle takes on our emotional, mental, and spiritual health. After the collapse, life will slow down, and we will appreciate things more, take less for granted, and become a more satisfied people.

TABLE 12.3 - GOOD FROM THE COLLAPSE
1. We will be freed from our materialism.
2. Life will be simpler and more satisfying.
3. God will be the center of our hopes and dreams.
4. The gospel may spread like wildfire again.

God will once again be the center of our hopes and dreams after the collapse. No longer will we be able to put our trust in our government, or our bank account, or our 401k plan. We will need God in a way we may have never needed him before, and he will become more real to us than he has ever been. Moreover, the empty "American Dream" that has occupied our time and attention for so long will be replaced by a dream of heaven and a longing for eternal life. Because of the collapse, it will be easier for us to "seek those things which are above, where Christ is, sitting at the right hand of God" and to "set [our] mind on things above, not on things on the earth" (Colossians 3:1-2).

The collapse will also make a way for the gospel to spread like wildfire. Can you imagine the hearts that will open to Christ when all of our wealth and pride are taken away? I've heard many preachers over the years cry out for another Great Awakening in America. It is quite possible that the collapse could bring that very thing. This judgment, if we look at it from the right perspective, is not something to dread—it is something to welcome. It will be a time for the church to grow beyond what we could imagine today, and for Christ to be glorified in our nation like never before.

LET US PRAY TOGETHER

As we close this book and our discussion about socialism, capitalism, and the future of American prosperity, let me offer a prayer for you, God's people, and our nation:

> Glorious God of the Bible, I bow my head now in the name of Jesus Christ, your Son. As much as I can, I offer a prayer of repentance for my nation and the churches within it. We have sinned against so much light, and we have thrown your blessings back in your face through our evil ways. Please forgive this great debt we have with you, and look upon us with mercy and not

FIGURE 12.7 - ANOTHER GREAT AWAKENING?

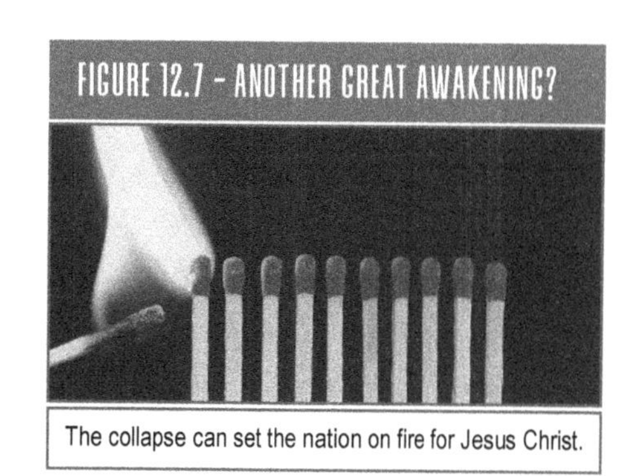

The collapse can set the nation on fire for Jesus Christ.

judgment.

Oh God, if you are willing, please stay your hand of justice. It is very clear that you are planning to lay a powerful blow against us in the days to come. We deserve every stroke of your rod. But we humbly ask that you withhold this punishment and instead grant us a great revival of faith in you in this land.

If you are not willing to do this, glorify yourself through the things you must do. Show the people of this nation, and the world at large, that you are a holy and righteous being who will not stand for constant sin. Demonstrate your overwhelming power by bringing low this nation, one of the greatest ever to exist on your earth.

And please help your people prepare for what is coming. Grant us to see the signs of the times and to understand the economic and financial principles that will lead us to the truth. Give us wisdom in terms of what we should do, and grant us protection when your sword is drawn upon our land. We look to you for our deliverance and survival. You are all we have.

Through this event, draw us to you in great faith, and purge us of the materialism, greed, and arrogance our wealth has brought us. Make us a people who truly please you.

We commit our lives and futures to your grace. Be ever merciful to us in Jesus name.

Amen.

CHAPTER DISCUSSION QUESTIONS

For a better understanding of the ideas in this chapter, work through these discussion questions in a small group:

1. What does God promise to do in the darkest times (like times of famine, violence, and war) for his faithful people? What verses teach this promise?
2. How can you make others aware of America's financial situation before it is too late?
3. What role does prayer have in preparing for the collapse?
4. Describe some of the individual prep-

arations reviewed in this chapter. Which do you think will be the hardest to do? What strategies can we use to accomplish it anyway?

5. What ministry does the chapter recommend for learning how to get out of debt? Had you heard of this ministry before? Explain your experience with it, if any.
6. How do the recommendations for church preparedness compare to those for individuals? Are there ones that are the same? Which?
7. Explain how churches should ban together to support one another during and after the collapse. Is this important? Why or why not?
8. What good can come to us through an event as devastating as the collapse?
9. How could the collapse make it possible for the gospel to spread like wildfire in the nation?
10. Do you think the faith of many will fail following the collapse? Why or why not? Will your faith fail? How can you prevent this from happening?

References:

Abrams, Margaret. "Barack and Michelle Obama net worth 2020: How much is the former US President worth along with his wife?" *Evening Standard*, 19 February 2020. https://www.standard.co.uk/insider/alist/barack-and-michelle-obama-net-worth-2020-how-much-is-the-former-us-president-worth-along-with-his-a4178561.html. Accessed 3 December 2020.

Amadeo, Kimberly, and Somer G. Anderson. "Who Owns the US National Debt?" *The Balance*, 14 October 2020. https://www.thebalance.com/who-owns-the-u-s-national-debt-3306124#:~:text=The%20public%20holds%20over%20%2421,insurance%20companies%2C%20and%20savings%20bonds. Accessed 5 December 2020.

Barua, Bacchus, and Mackenzie Moir. "Waiting Your Turn: Wait Times for Health Care in Canada, 2019 Report." *Fraser Institute*, 10 December 2019. https://www.fraserinstitute.org/studies/waiting-your-turn-wait-times-for-health-care-in-canada-2019#:~:text=Specialist%20physicians%20surveyed%20report%20a,19.8%20weeks%20reported%20in%202018. Accessed 5 December 2020.

Blahous, Charles. "The Fiscal Effects of Repealing the Affordable Care Act." *Mercatus Research*, 4 April 2017. https://papers.ssrn.com/sol3/papers.cfm?abstract_id=3211638. Accessed 4 December 2020.

Cain, Aine. "6 Incredible Perks for Apple Employees." *Inc*, 22 November 2017. https://www.inc.com/business-insider/apple-employee-benefits-perks-glassdoor.html. Accessed 4 December 2020.

Colestock, Stephanie. "18 Ways to Get Free Money from the Government." *DoughRoller*, 24 November 2020. https://www.doughroller.net/money-life/18-ways-to-get-free-money-from-the-government/. Accessed 4 December 2020.

Congressional Budget Office. "The Federal Budget in 2019: An Infographic." 15 April 2020. https://www.cbo.gov/publication/56324. Accessed 16 December 2020.

Congressional Budget Office. "Federal Debt: A Primer." March 2020. https://www.cbo.gov/publication/56309#:~:text=%2Fpublication%2F56020.-,Chapter%201%3A%20Debt%20Held%20by%20the%20Public,off%20debt%20as%20it%20matures. Accessed 5 December 2020.

Congressional Budget Office. "The Federal Budget in 2019: An Infographic." 15 April 2020. https://www.cbo.gov/publication/56324. Accessed 4 December 2020.

CSI Market. "S&P 500 Profitability." https://csimarket.com/Industry/industry_Profitability_Ratios.php?sp5. Accessed 4 December 2020.

Davies, Antony, Bruce Yandle, Derek Thieme, and Robert Sarvis. “The U.S. Experience with Fiscal Stimulus: A Historical and Statistical Analysis of U.S. Fiscal Stimulus Activity, 1953-2011.” *Mercatus Research*, April 2012. https://www.mercatus.org/system/files/US-Experience-Fiscal-Stimulus.pdf#page=7. Accessed 1 December 2020.

Davis, Wynne. “California Wildfires Have Burned 4 Million Acres And The Season Isn't Over Yet.” *NPR*, 4 October 2020. https://www.npr.org/2020/10/04/920154138/california-wildfires-have-burned-4-millions-acres-and-the-season-isnt-over-yet. Accessed 1 December 2020.

Debt.org. “ObamaCare and Associated Costs.” https://www.debt.org/medical/obamacare/. Accessed 4 December 2020.

Dornbusch, Rudiger, Stanley Fischer, and Richard Startz. *Macroeconomics*, 13th ed., New York, McGraw-Hill Education, 2018.

Dupor, Bill. “The Recovery Act of 2009 vs. FDR's New Deal: Which Was Bigger?” *Federal Reserve Bank of St. Louis*, 10 February 2017. https://www.stlouisfed.org/publications/regional-economist/first_quarter_2017/the-recovery-act-of-2009-vs-fdrs-new-deal-which-was-bigger. Accessed 5 December 2020.

Ewart, John. Level 1 Training DVD Course. Louisville, KY, Society for Church Consulting, 2008.

Fagan, Patrick. “How Broken Families Rob Children of Their Chances for Future Prosperity.” *The Heritage Foundation,* 11 June 1999. https://www.heritage.org/marriage-and-family/report/how-broken-families-rob-children-their-chances-future-prosperity. Accessed 15 December 2020.

Federal Reserve Bank of St. Louis. “Federal Debt: Total Public Debt.” https://fred.stlouisfed.org/series/GFDEBTN. Accessed 16 December 2020.

Federal Reserve Bank of St. Louis. “Federal Surplus or Deficit.” https://fred.stlouisfed.org/series/FYFSD. Accessed 5 December 2020.

Forbes. “ExxonMobil (XOM).” https://www.forbes.com/companies/exxon-mobil/#15fc8b99601f. Accessed 4 December 2020.

Forbes. “The Richest in 2020.” https://www.forbes.com/billionaires/. Accessed 3 December 2020.

Foussianes, Chole. “How Bernie Sanders Became a Millionaire.” *Town and Country Magazine*, 15 March 2020. https://www.townandcountrymag.com/society/politics/a31437248/bernie-sanders-net-worth/. Accessed 3 December 2020.

Fox Business. “China will overtake US as world's top economy in 2028, think tank says.” 25 December 2020. https://www.foxbusiness.com/economy/china-will-overtake-us-as-worlds-top-economy-in-2028-think-tank-says. Accessed 26 December 2020.

Government Accounting Office. "U.S. Postal Service's Financial Viability - High Risk Issue." https://www.gao.gov/key_issues/us_postal_service_financial_viability/issue_summary. Accessed 4 December 2020.

Granville, Brigitte. *Remembering Inflation.* E-book, Princeton, NJ, Princeton University Press, 2013). EBSCOhost, search.ebscohost.com/login.aspx?direct=true&db=nlebk&AN=571951&site=ehost-live. Accessed 2 December 2020.

Henney, Megan. "How much money is Nancy Pelosi worth?" *Fox Business*, 17 July 2020. https://www.foxbusiness.com/money/how-much-money-is-nancy-pelosi-worth. Accessed 3 December 2020.

Investopedia. "U.S. Bonds vs. Bills and Notes: What's the Difference?" 24 July 2020. https://www.investopedia.com/ask/answers/difference-between-bills-notes-and-bonds/. Accessed 5 December 2020.

Jones, Marc. "Exclusive: Second sovereign downgrade wave coming, major nations at risk - S&P Global." *Reuters*, 16 October 2020. https://www.reuters.com/article/us-global-ratings-sovereign-s-p-exclusiv/exclusive-second-sovereign-downgrade-wave-coming-major-nations-at-risk-sp-global-idUSKBN27126V. Accessed 7 December 2020.

Katz, Diane. "Here's how much red tape Trump has cut." *Heritage*, 17 October 2018. https://www.heritage.org/government-regulation/commentary/heres-how-much-red-tape-trump-has-cut." Accessed 5 December 2020.

Kelly, John, Jim Sergent, and Donovan Slack. "Death rates, bedsores, ER wait times: Where every VA hospital lags or leads other medical care." *USA Today*, 7 February 2019. https://www.usatoday.com/in-depth/news/investigations/2019/02/07/where-every-va-hospital-lags-leads-other-care/2511739002/. Accessed 4 December 2020.

Macrotrends. "Walmart: Number of Employees 2006-2020/WMT." https://www.macrotrends.net/stocks/charts/WMT/walmart/number-of-employees. Accessed 4 December 2020.

MarketWatch. "Exxon Mobil Corp." https://www.marketwatch.com/investing/stock/xom/financials/income/quarter. Accessed 4 December 2020.

McBride, William. "What is the evidence on taxes and growth?" *The Tax Foundation*, 18 December 2012. https://taxfoundation.org/what-evidence-taxes-and-growth/#:~:text=Particularly%2C%20they%20find%20that%20a,the%20main%20channel%20is%20investment. Accessed 7 December 2020.

McEachern, William A. *MacroEcon: Principles of Macroeconomics*, 6th ed. Boston, Cengage Learning, 2019.

National Archives. "Fact No. 1, [11 September 1792]." https://founders.archives.gov/documents/Hamilton/01-12-02-0274. Accessed 5 December 2020.

Newman, Katelyn. "Unforgettable U.S. Natural Disasters of the 2010s." *US News*, 27 December 2019. https://www.usnews.com/news/best-states/slideshows/the-10-most-unforgettable-us-natural-disasters-of-the-past-decade. Accessed 1 December 2020.

Nowrasteh, Alex, and Robert Orr. "Immigration and the Welfare State: Immigrant and Native Use Rates and Benefit Levels for Means-Tested Welfare and Entitlement Programs." *Cato Institute*, 10 May 2018. https://www.cato.org/publications/immigration-research-policy-brief/immigration-welfare-state-immigrant-native-use-rates. Accessed 4 December 2020.

Pariona, Amber. "Murder Rate by Country." *WorldAtlas*, 9 January 2020. https://www.worldatlas.com/articles/murder-rates-by-country.html. Accessed 1 December 2020.

Pham, Kevin. "America Outperforms Canada in Surgery Wait Times—And It's Not Even Close." *Foundation for Economic Education*, 17 July 2019. https://fee.org/articles/america-outperforms-canada-in-surgery-wait-times-and-its-not-even-close/. Accessed 5 December 2020.

Pugel, Thomas. *International economics*, 17th ed. E-book, New York, McGraw-Hill Education, 2020.

RaboResearch – Economic Research. "The Mexican 1982 debt crisis: Economic report." 19 September 2013. https://economics.rabobank.com/publications/2013/september/the-mexican-1982-debt-crisis/#:~:text=In%20August%201982%2C%20Mexico%20was%20not%20able%20to%20service%20its,debt%20payments%20to%20rise%20sharply. Accessed 3 December 2020.

Rich, Robert. "The Great Recession." *Federal Reserve History*. https://www.federalreservehistory.org/essays/great-recession-of-200709#:~:text=Beyond%20its%20duration%2C%20the%20Great,data%20as%20of%20October%202013). Accessed 7 December 2020.

Roberts, Roxanne. "Why does everybody suddenly hate billionaires? Because they've made it easy." *The Washington Post*, 13 March 2019. https://www.washingtonpost.com/lifestyle/style/why-does-everybody-suddenly-hate-billionaires-because-theyve-made-it-easy/2019/03/13/00e39056-3f6a-11e9-a0d3-1210e58a94cf_story.html. Accessed 3 December 2020.

Romig, Kathleen. "What the 2019 Trustees' Report Shows About Social Security." *Center on Budget and Policy Priorities*, 5 June 2019. https://www.cbpp.org/research/social-security/what-the-2019-trustees-report-shows-about-social-security. Accessed 8 December 2020.

Sabanoglu, Tugba. "Number of Amazon.com employees 2007-2019." *Statista*, 30 November 2020. https://www.statista.com/statistics/234488/number-of-amazon-employees/

#:~:text=In%202019%2C%20the%20American%20multinational,dollars%20in%202019%20net%20revenues. Accessed 4 December 2020.

Samenow, Jason Andrew Freedman, and Matthew Cappucci. "2020 Atlantic hurricane season breaks all-time record while leaving Gulf Coast battered." *Washington Post*, 10 November 2020. https://www.washingtonpost.com/weather/2020/11/10/record-hurricane-season-atlantic/. Accessed 1 December 2020.

Senate.gov. "Washington's Farewell Address to the People of the United States." https://www.senate.gov/artandhistory/history/resources/pdf/Washingtons_Farewell_Address.pdf. Accessed 5 December 2020.

Sorace, Stephen. "Whole Foods CEO Slams Socialism as 'Trickle-up Poverty': 'It doesn't work.'" *Fox Business*, 29 November 2020. https://www.foxbusiness.com/business-leaders/whole-foods-ceo-socialism-poverty-capitalism. Accessed 2 December 2020.

Tan, Weizhen. "The growing U.S. deficit raises questions about funding as China cuts U.S. debt holdings." *CNBC*, 2 November 2020. https://www.cnbc.com/2020/11/02/china-drops-us-treasurys-impact-on-us-deficit-and-coronavirus-stimulus.html. Accessed 5 December 2020.

Tanner, Michael, and Charles Hughes. "The Work vs. Welfare Trade-Off: An Analysis of the Total Level of Welfare Benefits by State." *Cato Institute*, 2013. https://www.cato.org/sites/cato.org/files/pubs/pdf/the_work_versus_welfare_trade-off_2013_wp.pdf. Accessed 3 December 2020.

The Jefferson Monticello. "Extract from Thomas Jefferson to 'Henry Tompkinson' (Samuel Kercheval)." http://tjrs.monticello.org/letter/1383. Accessed 5 December 2020.

Trading Economics. "United States – Credit Rating." https://tradingeconomics.com/united-states/rating#:~:text=Standard%20%26%20Poor's%20credit%20rating%20for,at%20Aaa%20with%20stable%20outlook. Accessed 7 December 2020.

U.S. Department of the Treasury. "The CARES Act Works for All Americans." https://home.treasury.gov/policy-issues/cares. Accessed 5 December 2020.

Williamson, Kevin. D. *The Politically Incorrect Guide to Socialism.* Washington, DC, Regnery, 2011.

World Population Review. "Murder Rate by Country 2020." https://worldpopulationreview.com/country-rankings/murder-rate-by-country. Accessed 21 December 2020.

Worstall, Tim. "The Average US Welfare Payment Puts You In The Top 20% Of All Income Earners." *Forbes*, 4 May 2015. https://www.forbes.com/sites/timworstall/2015/05/04/the-average-us-welfare-payment-puts-you-in-the-top-20-of-all-income-earners/#2c161cc316f0. Accessed 3 December 2020.

Yale Law School. "First Annual Message of John Adams." https://avalon.law.yale.edu/18th_century/adamsme1.asp. Accessed 5 December 2020.

York, Erica. "Summary of the Latest Federal Income Tax Data, 2020 Update." *Tax Foundation*, 25 February 2020. https://taxfoundation.org/summary-of-the-latest-federal-income-tax-data-2020-update/. Accessed 3 December 2020.

Ziol-Guest, Kathleen M., Greg J. Duncan, and Ariel Kalil. "One-Parent Students Leave School Earlier." *Education Next*, Spring 2015. https://www.educationnext.org/one-parent-students-leave-school-earlier/. Accessed 15 December 2020.

APPENDIX A

Justice David J. Bremer's Historical Analysis of America as a Christian Nation[1]
as appeared in the Supreme Court Opinion of the *Church of the Holy Trinity v. the United States*, February 29, 1892

CHURCH OF THE HOLY TRINITY v. UNITED STATES.
No. 143.
SUPREME COURT OF THE UNITED STATES
143 U.S. 457; 12 S. Ct. 511; 1892 U.S. LEXIS 2036; 36 L. Ed. 226

Argued and submitted January 7, 1892.
February 29, 1892, Decided

ERROR TO THE CIRCUIT COURT OF THE UNITED STATES FOR THE SOUTHERN DISTRICT OF NEW YORK.

The case is stated in the opinion.

SYLLABUS: The act of February 26, 1885, "to prohibit the importation and migration of foreigners and aliens under contract or agreement to perform labor in the United States, its Territories, and the District of Columbia," 23 Stat. 332, c. 164, does not apply to a contract between an alien, residing out of the United States, and a religious society incorporated under the laws of a State, whereby he engages to remove to the United States and to enter into the service of the society as its rector or minister.

COUNSEL: Mr. Seaman Miller for plaintiff in error.

Mr. Assistant Attorney General Maury for defendant in error submitted on his brief.

OPINION: MR. JUSTICE BREWER delivered the opinion of the court.

Plaintiff in error is a corporation, duly organized and incorporated as a religious society under the laws of the State of New York. E. Walpole Warren was, prior to Septem-

1. Copied word-for-word from the original opinion. Legal arguments intentionally removed for space. Long paragraphs broken for readability.

ber, 1887, an alien residing in England. In that month the plaintiff in error made a contract with him, by which he was to remove to the city of New York and enter into its service as rector and pastor; and in pursuance of such contract, Warren did so remove and enter upon such service. It is claimed by the United States that this contract on the part of the plaintiff in error was forbidden by the act of February 26, 1885, 23 Stat. 332, c. 164, and an action was commenced to recover the penalty prescribed by that act. The Circuit Court held that the contract was within the prohibition of the statute, and rendered judgment accordingly, (36 Fed. Rep. 303;) and the single question presented for our determination is whether it erred in that conclusion.

[LEGAL ARGUMENTS FOUND HERE REMOVED FOR SPACE.]

We find therefore, that the title of the act, the evil which was intended to be remedied, the circumstances surrounding the appeal to Congress, the reports of the committee of each house, all concur in affirming that the intent of Congress was simply to stay the influx of this cheap unskilled labor.

But beyond all these matters no purpose of action against religion can be imputed to any legislation, state or national, because this is a religious people. This is historically true. From the discovery of this continent to the present hour, there is a single voice making this affirmation. The commission to Christopher Columbus, prior to his sail westward, is from "Ferdinand and Isabella, by the grace of God, King and Queen of Castile," etc., and recites that "it is hoped that by God's assistance some of the continents and islands in the ocean will be discovered," etc. The first colonial grant, that made to Sir Walter Raleigh in 1584, was from "Elizabeth, by the grace of God, of England, [France][2] and Ireland, queen, defender of the faith," etc.; and the grant authorizing him to enact statutes for the government of the proposed colony provided that "they be not against the true Christian faith now professed in the Church of England."

[3]The first charter of Virginia, granted by King James I in 1606, after reciting the application of certain parties for a charter, commenced the grant in these words: "We, greatly commending, and graciously accepting of, their Desires for the Furtherance of so noble a Work, which may, by the Providence of Almighty God, hereafter tend to the Glory of his Divine Majesty, in propagating of Christian Religion to such People, as yet live in Darkness and miserable Ignorance of the true Knowledge and Worship of God, and may in time bring the Infidels and Savages, living in those parts, to human Civility, and to a settled and quiet Government; DO, by these our Letters-Patents, graciously accept of, and agree to, their humble and well-intended Desires."

---

2. Old spellings replaced with modern words in brackets.
3. Paragraph divided for readability.

Language of similar import may be found in the subsequent charters of that colony, from the same king, in 1609 and 1611; and the same is true of the various charters granted to the other colonies. In language more or less emphatic is the establishment of the Christian religion declared to be one of the purposes of the grant. The celebrated compact made by the Pilgrims in the Mayflower, 1620, recites: "Having undertaken for the Glory of God, and Advancement of the Christian Faith, and the Honour of our King and Country, a Voyage to plant the first Colony in the northern Parts of Virginia; Do by these Presents, solemnly and mutually, in the Presence of God and one another, covenant and combine ourselves together into a civil Body Politick, for our better Ordering and Preservation, and Furtherance of the Ends aforesaid."

The fundamental orders of Connecticut, under which a provisional government was instituted in 1638-1639, commence with this declaration: "Forasmuch as it hath pleased the [Almighty] God by the wise disposition of his [divine] prudence so to Order and dispose of things that we the Inhabitants and Residents of Windsor, Hartford and Wethersfield are now cohabiting and dwelling in and vapor the River of [Connecticut] and the Lands thereunto [adjoining]; And well knowing where a people are gathered together the word of God requires that to maintain the peace and union of such a people there should be an orderly and decent [Government] established according to God, to order and dispose of the affairs of the people at all seasons as occasion shall require; doe therefore associate and [conjoin] our [souls] to be as one [Public] State or [Commonwealth]; and do, for our [souls] and our Successors and such as shall be adjoined to us at any time hereafter, enter into Combination and Confederation together, to maintain and pressure the liberty and purity of the gospel of our Lord Jesus which we now profess, as also the discipline of the Churches, which according to the truth of the said gospel is now [practiced] amongst [us]."

In the charter of privileges granted by William Penn to the province of Pennsylvania, in 1701, it is recited: "Because no People can be truly happy, though under the greatest Enjoyment of Civil Liberties, if abridged of the Freedom of their Consciences, as to their Religious Profession and Worship; And Almighty God being the only Lord of Conscience, Father of Lights and Spirits; and the Author as well as Object of all divine Knowledge, Faith and Worship, who only doth enlighten the Minds, and persuade and convince the Understandings of People, I do hereby grant and declare," etc.

Coming nearer to the present time, the Declaration of Independence recognizes the presence of the Divine in human affairs in these words: "We hold these truths to be self-evident, that all men are created equal, that they are endowed by their Creator with certain unalienable Rights, that among these are Life, Liberty, and the pursuit of Happiness." "We, therefore, the Representatives of the United States of America, in General Congress, Assembled, appealing to the Supreme Judge of the world for the rectitude of our intentions, do, in the Name and by Authority of the good People of these Colonies, solemnly publish and declare," etc.; "And for the support of this Declaration, with a firm reliance on the Protection of Divine Providence, we mutually pledge to each other our Lives, our Fortunes, and our sacred Honor."

If we examine the constitutions of the various States we find in them a constant recognition of religious obligations. Every constitution of every one of the forty-four States[4] contains language which either directly or by clear implication recognizes a profound reverence for religion and an assumption that its influence in all human affairs is essential to the well being of the community. This recognition may be in the preamble, such as is found in the constitution of Illinois, 1870: "We, the people of the State of Illinois, grateful to Almighty God for the civil, political and religious liberty which He hath so long permitted us to enjoy, and looking to Him for a blessing upon our endeavors to secure and transmit the same unimpaired to succeeding generations," etc.

It may be only in the familiar requisition that all officers shall take an oath closing with the declaration "so help me God." It may be in clauses like that of the constitution of Indiana, 1816, Article XI, section 4: "The manner of administering an oath or affirmation shall be such as is most consistent with the conscience of the deponent, and shall be esteemed the most solemn appeal to God." Or in provisions such as are found in Articles 36 and 37 of the Declaration of Rights of the Constitution of Maryland, 1867:

> [5]That as it is the duty of every man to worship God in such manner as he thinks most acceptable to Him, all persons are equally entitled to protection in their religious liberty; wherefore, no person ought, by any law, to be molested in his person or estate on account of his religious persuasion or profession, or for his religious practice, unless, under the color to religion, he shall disturb the good order, peace or safety of the State, or shall infringe the laws of morality, or injure others in their natural, civil or religious rights; nor ought any person to be compelled to frequent or maintain or contribute, unless on contract, to maintain any place of worship, or any ministry; nor shall any person, otherwise competent, be deemed incompetent as a witness, or juror, on account of his religious belief:
>
> Provided, He believes in the existence of God, and that, under His dispensation, such person will be held morally accountable for his acts, and be rewarded or punished therefor, either in this world or the world to come. That no religious test ought ever to be required as a qualification for any office of profit or trust in this State other than a declaration of belief in the existence of God; nor shall the legislature prescribe any other oath of office than the oath prescribed by this constitution.

---

4. At the time of this court decision, there were 44 states in the United States.
5. Paragraph divided and block indent added.

Or like that in Articles 2 and 3, of Part 1st, of the Constitution of Massachusetts, 1780:

> [6]It is the right as well as the duty of all men in society publicly and at stated seasons, to worship the Supreme Being, the great Creator and Preserver of the universe.... As the happiness of a people and the good order and preservation of civil government essentially depend upon piety, religion and morality, and as these cannot be generally diffused through a community but by the institution of the public worship of God and of public instructions in piety, religion and morality:
>
> Therefore, to promote their happiness and to secure the good order and preservation of their government, the people of this commonwealth have a right to invest their legislature with power to authorize and require, and the legislature shall, from time to time, authorize and require, the several towns, parishes, precincts and other bodies politic or religious societies to make suitable provision, at their own expense, for the institution of the public worship of God and for the support and maintenance of public Protestant teachers of piety, religion and morality in all cases where such provision shall not be made voluntarily.

Or as in sections 5 and 14 of Article 7, of the constitution of Mississippi, 1832: "No person who denies the being of a God, or a future state of rewards and punishments, shall hold any office in the civil department of this State.... Religion, morality and knowledge being necessary to good government, the preservation of liberty, and the happiness of mankind, schools and the means of education, shall forever be encouraged in this State." Or by Article 22 of the constitution of Delaware, 1776, which required all officers, besides an oath of allegiance, to make and subscribe the following declaration: "I, A.B., do profess faith in God the Father, and in Jesus Christ His only Son, and in the Holy Ghost, one God, blessed for evermore; and I do acknowledge the Holy Scriptures of the Old and New Testament to be given by divine inspiration."

Even the Constitution of the United States, which is supposed to have little touch upon the private life of the individual, contains in the First Amendment a declaration common to the constitutions of all the States, as follows: "Congress shall make no law respecting an establishment of religion, or prohibiting the free exercise thereof," etc. And also provides in Article 1, section 7, (a provision common to many constitutions,) that the Executive shall have ten days (Sundays excepted) within which to determine whether he will approve or veto a bill.

There is no dissonance in these declarations. There is a universal language pervading them all, having one meaning; they affirm and reaffirm that this is a religious nation.

6. Paragraph divided and block indent added.

These are not individual sayings, declarations of private persons: they are organic utterances; they speak the voice of the entire people. While because of a general recognition of this truth the question has seldom been presented to the courts, yet we find that in *Updegraph v. The Commonwealth*, 11 S. & R. 394, 400, it was decided that, "Christianity, general Christianity, is, and always has been, a part of the common law of Pennsylvania; ... not Christianity with an established church, and tithes, and spiritual courts; but Christianity with liberty of conscience to all men." And in *The People v. Ruggles*, 8 Johns. 290, 294, 295, Chancellor Kent, the great commentator on American law, speaking as Chief Justice of the Supreme Court of New York, said:

> [7]The people of this State, in common with the people of this country, profess the general doctrines of Christianity, as the rule of their faith and practice; and to scandalize the author of these doctrines is not only, in a religious point of view, extremely impious, but, even in respect to the obligations due to society, is a gross violation of decency and good order... The free, equal and undisturbed enjoyment of religious opinion, whatever it may be, and free and decent discussions on any religious subject, is granted and recurred; but to revile, with malicious and blasphemous contempt, the religion professed by almost the whole community; is an abuse of that right. Nor are we bound, by any expressions in the Constitution as some have strangely supposed, either not to punish at all, or to punish indiscriminately, the like attacks upon the religion of Mahomet or of the Grand Lama; and for this plain reason, that the case assumes that we are a Christian people, and the morality of the country is deeply engrafted upon Christianity, and not upon the doctrines or worship of those impostors.

And in the famous case of Vidal v. Girard's Executors, 2 How. 127, 198, this court, while sustaining the will of Mr. Girard, with its provision for the creation of a college into which no minister should be permitted to enter, observed: "It is also said, and truly, that the Christian religion is a part of the common law of Pennsylvania."

If we pass beyond these matters to a view of American life as expressed by its laws, its business, its customs and its society, we find everywhere a clear recognition of the same truth. Among other matters note the following: The form of oath universally prevailing, concluding with an appeal to the Almighty; the custom of opening sessions of all deliberative bodies and most conventions with prayer; the prefatory words of all wills, "In the name of God, amen;" the laws respecting the observance of the Sabbath, with the general cessation of all secular business, and the closing of courts, legislatures, and other similar public assemblies on that day; the churches and church organizations which abound in every city, town and hamlet; the multitude of charitable organizations existing every

7. Paragraph divided and block indent added.

where under Christian auspices; the gigantic missionary associations, with general support, and aiming to establish Christian missions in every quarter of the globe. These, and many other matters which might be noticed, add a volume of unofficial declarations to the mass of organic utterances that this is a Christian nation. In the face of all these, shall it be believed that a Congress of the United States intended to make it a misdemeanor for a church of this country to contract for the services of a Christian minister residing in another nation?

Suppose in the Congress that passed this act some member had offered a bill which in terms declared that, if any Roman Catholic church in this country should contract with Cardinal Manning to come to this country and enter into its service as pastor and priest; or any Episcopal church should enter into a like contract with Canon Farrar; or any Baptist church should make similar arrangements with Rev. Mr. Spurgeon; or any Jewish synagogue with some eminent Rabbi, such contract should be adjudged unlawful and void, and the church making it be subject to prosecution and punishment, can it be believed that it would have received a minute of approving thought or a single vote? Yet it is contended that such was in effect the meaning of this statute.

[8]The construction invoked cannot be accepted as correct. It is a case where there was presented a definite evil, in view of which the legislature used general terms with the purpose of reaching all phases of that evil, and thereafter, unexpectedly, it is developed that the general language thus employed is broad enough to reach cases and acts which the whole history and life of the country affirm could not have been intentionally legislated against. It is the duty of the courts, under those circumstances, to say that, however broad the language of the statute may be, the act, although within the letter, is not within the intention of the legislature, and therefore cannot be within the statute.

The judgment will be reversed, and the case remanded for further proceedings in accordance with this opinion.

8. Paragraph divided for readability.

APPENDIX B

The Basics of Suffering
From *Basic Bible Christianity*[1]
by Wes Moore

Why does God let us suffer? Of all the questions asked about God, the Bible, or Christianity, this is by far the toughest to answer. Why? Because it causes two basic yet extremely important truths to come into conflict. The first is that God is good and powerful. If there is a God, whatever else he is, he is good, and he is powerful. Everybody knows this. The second is that life is full of misery and death. There is no doubt about this either, for every human being who has ever lived has experienced the pain and heartache of this life firsthand.

The problem comes when we try to bring these two truths together, to understand how they can both be true at the same time. How can God be good and powerful, and yet there still be so much suffering in the world? Making sense of this is nearly impossible for most people. As a result, millions refuse to believe there is a God at all, and millions of others live in doubt as to his goodness and love. But is this really necessary? Is there no answer to the question of suffering?

I'm here to tell you that there is. The Bible provides a complete and thorough answer that maintains both the goodness and power of God, and the existence of evil, death, and suffering. In this chapter, I want to share that answer with you. I'll do this by addressing the issue in three phases. First, I'll explain why there is suffering in the first place; second, I'll show you how God uses it for his good purposes; and, third, I'll share how you can turn your suffering into a blessing for yourself and others. Let us begin, then, with why suffering exists.

Why is there suffering?

In chapter 3 of *Basic Bible Christianity*, I explained the fall of man and the consequences of that fall. There I pointed out that God, the Creator, placed Adam in the garden of Eden and gave him a single, simple command: "Do not eat of this tree or you will die." Adam disobeyed this command, and God placed the entire creation—man, animals, and nature—under a curse of death. From that day forward, sickness, depression, murder, floods, hurricanes, famine, and all other forms of death and suffering became commonplace in the world. Death, then, is a result of the sin of man. We die because we sin.

But what is the relationship between sin and death? If we accept the Bible's teaching

---

1. All of Wes' books can be found at www.wesmoorenow.com. Free e-book copies are also available at this site.

on the connection between these two things, how do we make sense of it? In other words, why does death *have to come* because of sin (apart from the fact that God said it would)? The answer lies in a deeper understanding of our relationship to God.

God is our source of life; we only have life to the extent we are connected properly to him. In Acts 17:28, Paul made a very interesting statement on this point. He said, "For in Him [that is, in God] we live and move and have our being." In other words, our existence is dependent upon God. The earth we stand on, the sun we share, the breath we take, the bodies we have—they are held together and kept alive by him. Without his life, we cannot live.

Sin changed the relationship between all of creation and its source of life, God. When Adam rejected God's command, what he actually did was reject God himself. What Adam said was, in effect, "I don't want you around, God. Go away and stay away!" And, because he represented all of mankind, he didn't just say this for himself—he said it for all of us, every human being who would ever live. In doing so, he pushed our source of life away and opened the door for death to enter.

Let me picture this for you in another way. Consider the relationship between the sun and the earth. In a sense, the earth gets its life from the sun. The sun provides heat, light, and energy for all life on earth. The sun is not God, mind you, but in God's design, it is, in a sense, a source of life. Because of this life-giving relationship, if the earth was moved farther away from the sun, what would happen to life on earth? It would begin to experience more death, wouldn't it? And the farther it moved away, the more death it would experience. Then, at some point, if it was pushed far enough away, complete death would consume the planet. Nothing would survive.

This is what has happened to us in a spiritual sense. We have rejected God through sin, and therefore removed ourselves from our source of life, allowing death to enter in. Death, then, is our fault, not God's. We can raise our fists at him all we want for the suffering and death we see every day, but when all is said and done, we have only ourselves to blame.

However, because of God's love and mercy, death has not completely consumed us. Even though we operate under the curse, we still live, work, and, to some extent, experience happiness. But all the while, death surrounds us.

Yes, sin has pushed us away from our source of life causing death to enter. However, it is important to point out that death is only temporary in God's design. It will not go on forever. After the Final Judgment, those who have repented and trusted in Christ for salvation will be placed back into a perfect relationship with God and experience eternal life; their separation from God will end forever. On the other hand, those who reject Christ and the God of the Bible will be forced to pay for their sins in hell. This punishment is, in effect, being removed completely (not partially, as now) from God's life and suffering eternal death in hell. The subject of Final Judgment, eternal life, and hell will be covered in detail in chapter 12.

How does God use suffering to do good?

Now that we have a clearer understanding of why suffering exists, let's turn our attention to the ways God uses it to serve his greater purposes and to bring about good in our lives. The Bible gives at least five purposes of God in suffering. Let me explain them now.

Purpose 1: The punishment of evil acts. God uses suffering to punish the evil of man. While he holds his full judgment of sin until the Day of Judgment, he still carries out judgment in a smaller way now. This principle is called the "sowing and reaping principle" and is based on Galatians 6:7. In that passage, Paul says, "Do not be deceived, God is not mocked; for whatever a man sows, that he will also reap." What this means is that God does not let men get away with everything. He watches over their dirty deeds and uses his power to make them pay for what they do. When you hear that an evil dictator has been hung in the street, an adulterer has been murdered by a jealous husband, or a greedy preacher has lost his house to foreclosure, you are seeing this use of suffering in action.

Purpose 2: To drive the lost to repentance and faith. God also uses suffering to bring lost men and women to himself. As a race, we are so proud and arrogant. We think we've got it all figured out and have no need of God in our lives. In order to bring us to repentance and faith, the Lord must break us. He does this by bringing trial and difficulty into our lives. The Bible is full of examples of this, and so is our modern world. I came to Christ myself after he put me through some of the most difficult experiences of my life. If you are in the middle of a trial now and are not saved, he may be doing the same to you.

Purpose 3: To perfect the saved. Once we become Christians, suffering continues to play an important role in our lives. As mentioned in chapter 7, God's plan for his children is not just to save them but also to make them into the image of his Son, Jesus Christ. Suffering is his most effective tool for bringing this about. I will devote an entire section to this use of suffering later in this chapter. For now, let me simply share a verse that teaches this truth.

The author of Hebrews said, "For whom the Lord loves He chastens, and scourges every son whom He receives" (12:6). Here we are told that God "scourges" every one of his children. In ancient times, a scourging was a brutal beating used to punish criminals, and God uses it as a picture of what he does to us. This does not mean he will show up at our house and physically beat us, however. It does mean, though, that he will bring very difficult circumstances into our lives to train us to do good and keep us from doing evil. The writer confirms this a few verses later when he says, "For [our earthly fathers] indeed for a few days chastened [punished] us as seemed best to them, but [God does it] for our profit, that we may be partakers of His holiness" (Hebrews 12:10). In other words, the Lord sends trials our way to perfect us, to force us to change so we can become more like him.

Purpose 4: To show mercy to specific individuals. Sometimes we wonder why God would allow one of his children to die before their time. Maybe there's a woman who serves him daily by working among the poor. She has given up everything to obey his command to help the needy, and then one day she's hit by a car and dies at the ripe old age of 30. Why would God allow this? Similarly, why would he allow an unborn baby to be aborted or a child to drown in a swimming pool? The Bible tells us why he allows these things—to take these precious souls away from the suffering and death of this world.

Isaiah 57:1 says, "The righteous perishes, and no man takes it to heart; merciful men are taken away, while no one considers that the righteous is taken away from evil." The word *evil* here applies to the miseries, trials, and pain of this world. When God takes a believer or child before their time, he is showing them mercy by letting them skip the suffering and pain they would have had to endure if they had continued living in this sin cursed world.

Purpose 5: To glorify himself through martyrs. Some believers are chosen to accomplish a special act of service to Christ. The world is so angry with God that it would kill every believer if it could. Sometimes, to show his glorious power and grace, God will allow his disciples to suffer death for his name. Both Lazarus and Stephen from the New Testament are examples of this.

In John 11, Lazarus was allowed to die so Jesus could resurrect him and show the world his power. When Jesus was told of Lazarus' illness, he said, "This sickness is not unto death, but for the glory of God, that the Son of God may be glorified through it." A short while later, Lazarus was resurrected and many believed.

In Acts 7, Stephen preached a powerful sermon that led many to faith in Christ. The bad news—he was stoned to death for it. Verses 59-60 of that chapter describe how it happened, "And they stoned Stephen as he was calling on God and saying, 'Lord Jesus, receive my spirit.' Then he knelt down and cried out with a loud voice, 'Lord, do not charge them with this sin.' And when he had said this, he fell asleep [died]." God glorified himself by using Stephen's gracious attitude while he was being martyred to bring people to faith in Jesus, including the apostle Paul (58). He continues to use martyrdom in this way today.

God's Sovereignty Over Your Trial

One of the most important truths a Christian can learn about God is that he is sovereign over every event in the universe. The word *sovereign* (pronounced *sov'-er-in*) is not used very often today, but years ago it was a common term used to describe a king. A king was said to be a sovereign because he had complete control over everything that happened in his kingdom. He was the boss, the man, the top dog. Nothing happened without his permission, and everyone within his territory was subject to him.

God is also a king, a sovereign, but he does not just rule over a single nation or land; he is sovereign over the entire universe, every living thing, and all of time. Nothing—not even a speck of dust—escapes his full control. There are so many verses about this in the Bible, ten pages could not list them all. Here are a few of the more powerful ones for your review now:

> Yours, O LORD, is the greatness, the power and the glory, the victory and the majesty; for all that is in heaven and in earth is Yours; Yours is the kingdom, O LORD, and You are exalted as head over all. Both riches and honor come from You, and You reign over all. In Your hand is power and might; in Your hand it is to make great and to give strength to all. **1 Chronicles 29:11-12**
>
> All the inhabitants of the earth are reputed as nothing; He does according to His will in the army of heaven and among the inhabitants of the earth. No one can restrain His hand or say to Him, "What have You done?" **Daniel 4:35**
>
> Now see that I, even I, am He, and there is no God besides Me; I kill and I make alive; I wound and I heal; nor is there any who can deliver from My hand. **Deuteronomy 32:39**
>
> I know that You can do everything, and that no purpose of Yours can be withheld from You. **Job 42:2**
>
> Who is he who speaks and it comes to pass, when the LORD has not commanded it? **Lamentations 3:37**
>
> The LORD of hosts has sworn, saying, "Surely, as I have thought, so it shall come to pass, and as I have purposed, so it shall stand." **Isaiah 14:24**

Exactly what does this have to do with suffering? Everything, really. Because of God's power, nothing that happens to you happens without his approval. No matter how bad it is, it is always under his control. He is so powerful, in fact, it could be said that *nothing can happen to you unless it is God's will*. Does that frighten you? It shouldn't, because, as we have learned already, God is good, and he loves you. Therefore, everything bad that ever happens to you is sent by him to bring about a greater good in your life.

For some, this is a hard teaching to understand and an even harder one to accept. Could God actually be behind *all* the trials of my life? Does the Bible actually teach this? It absolutely does, and it teaches it everywhere.

Joseph is a famous Israelite from the Bible. He was one of the sons of Jacob (whom God renamed Israel), and was responsible for delivering his entire family, and the whole world, in fact, from a terrible famine in the ancient Middle East. As great a man as he was, the first half of his life was pure misery. He was sold as a slave by his brothers, falsely accused of attempted rape, and sent to prison for years without cause.

When his difficulties were finally over and he had had an opportunity to gain some perspective on his suffering, he said this to his brothers (the ones who had sold him as a slave): "But as for you, you meant evil against me; *but God meant it for good,* in order to bring it about as it is this day, to save many people alive. *It was not you who sent me here, but God*" (45:7 and 50:20). After many years to think about what had happened to him, Joseph saw that it was not ultimately his brothers who had sent him on the long path of trial—it was God. And he had done it to bring about a good for Joseph, his family, and countless others that no one could have understood at the time. Almost 1,500 years later, Paul would restate Joseph's words like this: "And we know that all things work together for good to those who love God, to those who are the called according to His purpose" (Romans 8:28).

When fully understood, this truth will completely change the way you look at suffering. You will begin to see it as God sees it, as his tool to bring about good in your life. Having this new understanding, you can move from always trying to get out of your trials (and complaining to God because of them) to finding his purpose in your difficulties and growing through the pain they bring.

Why Suffering is Necessary to Grow

Why does God use suffering to change us? Why can't he use gentler or easier ways to make us into the people he wants us to be? The answer has to do with our nature as human beings. You see, because of our sinful nature, if God does not make us uncomfortable, we will never change. When things are going well (from our perspective), we will keep doing what we have always done, even if those things are harmful to us, others, or the glory of God. For this reason, the Lord brings difficulties into our lives on purpose.

At the end of the day, don't you want to be happy, content, and at peace in your life, to rid yourself of worry, fear, and frustration? I would bet there's not a single person reading this paragraph who doesn't want that. Well, I've got good news for you: God wants these things for you too. The difference is in how he chooses to bring them about in your life.

Human beings try to create happiness by changing what's on the outside. When our wives don't make us happy, we look for a new one; when our jobs become too difficult, we move on to the next; when we don't feel good about ourselves, we lose weight, buy a new car, whiten our teeth—whatever it takes to make us feel better. Occasionally, we even get things the way we want on the outside, and for a while, we are happy. But that never lasts. Outside circumstances are always changing, and the peace the new car gave us is gone long before the shiny paint begins to fade.

Don't get me wrong: God gives us some of the things we want on the outside (many of them, in fact). But the primary way he makes us happy is to change us on the inside. Instead of giving us a new wife, he teaches us how to love the flawed one we have; instead of giving us a new job, he teaches us how to get along with the difficult people we work with; instead of giving us a smaller waistline, a new car, or shiny, white teeth, he shows us where our true value lies—in his love for us. By doing it this way, he creates a person who is happy, content, and at peace *no matter what the outside circumstances may be.*

Suffering Reveals Deeper Issues

So how does suffering help us grow? Suffering makes growth possible by showing us the deeper spiritual problems that need to be corrected. The trial itself is usually not the real problem. The real problem is something below the surface, something the trial itself is revealing.

As far as our discussion is concerned, trials can take one of two forms. First of all, they can show up as a sin in our lives that we can't get rid of. You may struggle with anger problems, for example. Whenever something doesn't go your way, you go off on someone, yell and scream, or punch a hole in the wall. This recurring sin in your life is itself a trial. Every time you do it, it causes you sadness, pain, and worry. Any kind of sin can serve as this type of sin trial.

Trials are not just a result of the sins we commit, though. Some have nothing to do with our particular sin problems at all (though all trials ultimately stem from the fall of man in the garden of Eden). They are simply events or situations that come up in our lives that bring hardship upon us. Take someone who has suffered with back pain for years as an example. The back problem itself is not a sin problem, but it still brings great difficulty into the person's life. Things like job loss, sickness, or financial problems also fit into this event trial category. It is also possible for good things to become sources of this kind of trial, like a new job or a new baby, because they too bring new challenges into our lives.

Whatever type of trial we face, we must look under the surface to find the issues God wants to change in us. Anger problems often come from a failure to forgive or a feeling of entitlement (that the world owes you something), and long-term health problems, like back issues, often reveal weaknesses in trust in God, complaining, and self-centeredness. Regardless of the problem on the surface, deeper issues are always at work.

The Lord will use both types of trials to reveal and correct spiritual issues within you. He may allow you (not *cause* you—you are always responsible for your choices) to cheat on your wife to teach you about faithfulness or resisting temptation; he may allow you to lose your job to show you how weak your faith is or how materialistic you have become; or he may allow deep-seated pain from childhood abuse to harm a relationship you value in order to train you how to forgive. Whatever the case, to grow from suffering, you must focus on finding and addressing the deeper issues the suffering reveals.

How to Find the Deeper Issues in Suffering

How do I know what my deeper issues are? The key is to keep asking questions about your trial until the issues become clear to you.

You should always begin by asking, "What is bothering me?" or "What is the problem?" Here you are simply trying to identify the source of the trial itself. The next question is, "Why is this bothering me?" or "Why is this a problem?" These types of questions take you a layer deeper and allow you to start analyzing your trial. You should continue asking "Why?" until you find the true spiritual issue (or issues). When an issue has finally been identified, you ask, "What spiritual weakness or need does this reveal in me?" Once this has been answered, you seek correction from the Bible by asking, "What does the Bible say about this?"

Let me give you an example from a conversation I had with a young married man recently. I asked him how he was doing (a first-level type of question), and he said, "I'm really stressed right now." Immediately, I recognized this as a trial the Lord had sent to help him grow. I then asked a second-level question, "Why are you stressed out?" He mentioned several things that were causing him stress, one of which was that the laundry at home was not getting done. I then asked, "Why is the laundry not getting done?" (Remember, keep asking "Why?" until you get to a true spiritual issue.) He said, "I work 60 hours a week, but my wife only works 35. I asked her to do it, but she said, 'You don't do it, so why should I?'" Although he didn't know it at the time, this answer revealed some of the spiritual issues at work in his trial. I then asked a third-level question: "What does this situation reveal about you or your wife from a spiritual standpoint?" He thought for a moment but couldn't give an answer.

I went on to explain to him that there were at least two spiritual issues that needed addressing here: one, his wife was not submitting to his authority as her husband, and, two, he was not training his wife in the Word as he was commanded. We then walked through the Bible's teachings on these issues and what he should do moving forward (a level four issue).

A few weeks later, I saw him again and asked how things were going with the laundry. He smiled and said the issue had been taken care of. "How did you do it?" I asked. "I just did what you told me to do from the Bible," he replied. (Mind you, not all issues are resolved this easily. Some take many years. However, the process is always the same.)

The point here is that the Lord allowed a situation to develop in this young man's life in order to reveal spiritual issues in him and his wife. To be clear, the specific questions will vary with each person and situation, but the overall process remains the same—ask questions to reveal deeper issues that need to be addressed.

Of course, you may need some help doing this at first. This is what pastors and wise, mature Christian brothers and sisters are for. They can help you not only understand the root issues Christ is showing you but also take you to places in the Bible where you can learn how to fix them.

APPENDIX C

The Basics of Salvation
From *Basic Bible Christianity*[1]
by Wes Moore

Job (pronounced *jobe*, like *robe*) is one of my favorite men in the Bible. Over the course of several years, he suffered more than anyone I have ever known. He lost his children, his wealth, his servants, and his health. Things got so bad at one point that he even cursed the day he was born. "Why did I not die at birth?" he asked in Job 3:11. "Why did I not perish when I came from the womb?"

After years of studying his book, I finally realize why God allowed him to suffer as he did, and it wasn't to punish him for some sin he had committed or to make him a punching bag for divine power. In the end, God put Job through all he did so he would have to wrestle with the deepest and most difficult questions a man could ever ask. Among other things, his pain made him wonder about suffering and about justice; in the face of so much loss, it is easy to see why. "How is it fair that all of this is happening to me?" Job might have asked.

But his experiences also caused him to wonder about something else, something perhaps even deeper and harder to answer—the way of salvation itself. The harsh realities of this life made Job think more about the life that is to come. And those thoughts led him to a realization of his own unworthiness before his Maker. In Job 9:2, he pondered, "But how can a man be righteous before God?" Indeed, Job, how can a man be right with God?

If you will take a moment to think about it, you will find that there is no more important question for any of us than this. How can *I* be righteous before God? How can I—a sinner by nature and a million dirty deeds—cause the Holy Creator to find me righteous? How could he ever see me as pure in his eyes and grant me salvation?

On the pages that follow, I want to take up this subject and show you—yes, *you*—exactly how you can be right with God and find the way of salvation. Before we dive into this critical subject, however, we must go back and review some of the things we learned in previous chapters.

What you must understand before you can be saved

Before outlining the steps in the salvation process (what we call "the gospel"), we need to highlight three things we have discussed thus far: God's holiness, man's evil na-

1. All of Wes' books can be found at www.wesmoorenow.com. Free e-book copies are also available at this site.

ture, and Christ's sacrifice for sin. These truths will provide the foundation upon which the plan of salvation can be built. In fact, without a clear understanding of these concepts, a person actually cannot be saved. Let's review them briefly now.

Foundational Truth 1: God is holy. From a right and wrong standpoint, God is holy. He never does anything wrong, never says anything wrong, never thinks anything wrong. He never sins, is never tempted to sin, and, in fact, cannot sin.

His perfect holiness means he is also perfectly just. He must punish everyone who sins, every time. He cannot just overlook sin as if it never happened. And, because of his sense of justice, the punishment he gives must always fit the crime. In the case of sin against him, the crime is so great the punishment must be eternal death in hell. No lesser penalty will do.

Foundational Truth 2: Man is evil. Unlike God, man is not holy. At his very core, he is evil. He always chooses to sin, is uninterested in a relationship with God, and views God as his enemy. He is completely rotten in his Creator's eyes and, no matter how hard he tries, cannot save himself.

These first two truths mean that every man and woman who has ever lived—that's all of us, by the way—has an eternal death sentence hanging over them. If something isn't done about our sin, when the Day of Judgment comes, we all will be sent to hell to pay for our crimes against God. It has already been determined; there is no way around it.

What, then, can be done? Is there no way to pay for sin without going to hell? Yes, there is, and that leads us to our final foundational truth.

Foundational Truth 3: Jesus died for sinners. Thankfully, God is also loving and merciful, and does not want us to suffer forever for our disobedience. For this reason, he sent Jesus Christ to live a perfect life for us and to die on the cross to pay for our sins. Because of his love for humanity, Jesus willingly obeyed the Father and sacrificed himself so we could be saved. This is why he is called "the Savior of the world" (John 4:42).

Now that we have reviewed these basic truths, the plan of salvation can be outlined.

How to be Saved

Taking advantage of God's mercy through the work of Jesus Christ requires a sinner to do three things. First, he must repent, second, he must put his faith in Christ, and, third, he must commit to serve Christ for the rest of his life. Let's take these in order.

repent of your sins

The first step to being saved is to repent of your sins. Repentance is more than just saying you're sorry, though. The kind of repentance the Bible talks about has several parts. Paul mentioned all of them in Acts 26:20. There he said that men "should repent, turn to God, and do works befitting repentance."

He begins with a call for men to repent. This word in the original language means to change your mind about sin and to begin to hate it. Repentance begins as a sadness in the heart over the sins you have committed. Eventually, that sadness makes you to want to stop sinning and leave those sins behind forever. Repenting means you see the pain and harm your sins have caused—in your own life, the lives of others, and in the heart of God—and, because of that pain, decide to turn away from them.

For true repentance to take place, the sinner must not only turn away from sin, he must also turn toward God for forgiveness of his sins. As Paul said, men "should repent [and] *turn to God.*" When we see what sin has done to us and others, and change our minds about it (the first step), we must then turn to God and ask for forgiveness for those sins.

The word *forgive* means "to release without payment." The word was originally a banking term used when a debt no longer had to be repaid. Let's say you owed a bank $10,000, and the bank president decided you no longer had to repay it. He would then forgive the debt, that is, he would release you from the need to pay it back. In the end, you would owe nothing.

In a similar way, when we sin against someone, we incur a debt to them. Our sin takes something from them that was not rightfully ours. Let's say I stole your car. Not only did I take the actual car from you, I also took your peace of mind (among other things). My debt to you for this sin would include all of this. If I ask for your forgiveness and you grant it to me, you are releasing me from the entire debt of the sin of stealing your car. You are, in effect, agreeing to take the loss yourself so I can go free.

When we ask God for forgiveness, we are asking that he release us from the sin debt we have with him without our paying it (through punishment in hell). When we do this, the Bible says he throws our debt out the window, along with the punishment that was due to us because of it. We are then free and clear, like the debt never even existed.

The sins do not really go unpunished, though; forgiveness is granted because Jesus died and paid for those sins. As with the stolen car example, God takes the loss himself by punishing his own son instead of us. More about this when we talk about faith in Christ later in the chapter.

The last part of repentance that Paul mentions is to "do works befitting [or, that prove] repentance." In other words, you must show your repentance is real. It is easy enough to say you have changed your mind about sin and to ask God for forgiveness. But God only grants forgiveness when we are sincere in our repentance. Basically what Paul is saying here is, "I am glad you have offered repentance to God, but you cannot be sure of forgiveness until your repentance is proven by your future conduct." For repentance to be real and forgiveness to be granted, the desire to repent must produce real change in a person's life.

place your faith in Christ

After true repentance is offered, the second step of salvation is to put your faith in Christ. But, someone might ask, why do I need to put my faith in Christ if God has already forgiven my sins when I repented? Let me use the example of a checking account to explain why.

Let's say I open a checking account with a balance of zero and have $1,000 of deductions on the first day. What would my balance be at the start of day two? Negative $1,000, right? Yes. After the first day, I would be in the hole by a grand. Now, if the bank forgave that negative balance (basically just forgot about it), what would my balance be then? Yes, zero. I would be out of the hole, but I would not have a positive balance in my account. In other words, I wouldn't have any money available to spend.

What does this have to do with salvation? A lot, actually. In God's eyes, because of our sins, we all have a negative account balance. However, our balance in more like a negative $1 billion than a measly $1,000. When we ask for forgiveness, he erases the negative balance in our account, bringing our balance back to zero (remember, this happens because Jesus died to pay that sin debt). However, to inherit eternal life, we must have money in our account; we must have a positive balance of righteousness to call our own. The problem is, forgiveness wipes away my negative balance, but it does not provide the positive righteousness I need to be saved.

For a person to be saved, he must also be declared righteous by God (have money in his account). God must look at him and see not only that he is forgiven but also that he is righteous. When God sees a person this way, the Bible calls it *justification*.

Justification is a legal act whereby God declares a sinner to be, not simply forgiven, but also righteous in his sight. That righteousness cannot come from the sinner, though, for we have already seen that no human being is righteous on his own (chapter 3). If we are going to have any, God must provide it for us, and he has done that through his son Jesus Christ.

Christ earned all of his righteousness during his perfect life on earth. He obeyed every command of the Father for all of his 33 years, including the command to offer himself on the cross as a sacrifice for humanity. As a result, he has built up an account balance of righteousness greater than all of the sins of everyone who has ever lived. His balance, in fact, is infinite.

While God forgives us for our sins when we repent, he only transfers the righteousness of Christ to our account when we put our faith in him. Faith in Christ becomes a pipe or conduit, so to speak, by which God sends us the righteousness of Jesus. So, then, what does it mean to put your faith in Christ?

Faith is not simply saying you believe that Jesus lived, or even that you agree with his teachings. It is a personal transfer of trust to him. When I say I believe in Jesus Christ, what I am saying is that I have fully transferred my trust for forgiveness and justification to his perfect life and sacrificial death for me.

A great way to understand saving faith is to imagine a man trapped in a small boat on the ocean during a storm. While he struggles to stay afloat on the violent waves, a helicopter drops down from above and lowers a rope that will pull him to safety. At that moment, the man has one of two choices: He can stay in the boat or he can grab the rope, but he cannot do both. In order to be saved from the storm, he must transfer his trust from the boat to the rope (and grab hold of it!). When he does, he is saying, "I give up all possibility of this boat saving me and put all of my trust in the rope. If I am wrong, I will die, for the boat will be gone, and I will drown."

When we put our trust in Christ for salvation, we are saying the same thing: "I no longer trust that I am good enough to earn God's forgiveness or that my good deeds can give me the righteousness I need. I completely transfer my trust to Jesus Christ. If his death does not count for me, and if his righteousness does not become mine, I will be lost." Anything short of this is not faith as the Bible defines it and will not result in salvation from sin or the granting of eternal life.

commit to Christ's service

Now that the basics of repentance and faith have been reviewed, it is time to explain the third step of salvation—committing to Christ's service. Jesus talked about this in detail in Luke 14:27-33:

> **27** And whoever does not bear his cross and come after Me cannot be My disciple.
>
> **28** For which of you, intending to build a tower, does not sit down first and count the cost, whether he has enough to finish it—
>
> **29** Lest, after he has laid the foundation, and is not able to finish, all who see it begin to mock him,
>
> **30** Saying, "This man began to build and was not able to finish."
>
> **31** Or what king, going to make war against another king, does not sit down first and consider whether he is able with ten thousand to meet him who comes against him with twenty thousand?
>
> **32** Or else, while the other is still a great way off, he sends a delegation and asks conditions of peace.
>
> **33** So likewise, whoever of you does not forsake all that he has cannot be My disciple.

The first thing to see here is that there is a life and death commitment required from every person Jesus saves. This can be seen in several ways. In verse 27, Jesus uses the imagery of a cross, an instrument of death, to show the commitment he demands. He is in effect saying, "If you don't have a cross, you can't be a disciple." In other words, "If you are not willing to *die* to follow me, you *cannot* follow me."

It can also be seen in the comparisons he uses to make his point, the king going to war and the man building the tower. The builder must have enough material to finish the job, not just start the project (28-30), and the king must have enough men to win the battle, not just start the fight (31-32). In each case, there is a finality implied: finish and win or give up and lose. "You've got to take this thing all the way to the end," Jesus is saying to those who would believe in him. "Therefore, consider what I am asking you to give up—everything." To remove all doubt as to his meaning, he finishes with the statement, "So likewise, whoever does not forsake [abandon or give up] all that he has cannot be My disciple" (33).

Another important thing to see in these verses is that this commitment is required *before* a person can become a disciple (before he can be saved, in other words). Notice how Jesus asks which builder or king "does not sit down *first*" and figure out if he can take his plan all the way (verses 28 and 31)? The fact is, Jesus doesn't offer a try-it-before-you-buy-it program as some car dealers do. You can't get on board the Salvation Express with a ticket just to the next station; there is only one ticket sold for this train, and it is for the full ride.

What does this mean for those who are considering following Jesus and being saved? It means you had better understand what he is asking of you before you believe. He wants your entire life, not just your name on a salvation card. (This doesn't mean you have to actually give up everything; it's more about the attitude of your heart, that you would be willing to if he asked you.)

It is costly to believe in Jesus Christ, so consider this carefully before you step out to follow him. However, remember this: There is nothing more joyful, satisfying, or rewarding in this life than dying to follow Jesus. I can tell you from personal experience—it's the best deal out here, hands down.

What happens when I believe?

As we come to the end of our discussion about how to be saved, I thought it would be helpful to briefly review all that happens when you complete the steps of salvation. Some of this we have talked about earlier in this chapter and some of it is new.

When I am saved, four important things happen. First, I am forgiven. As has already been said, this means that God wipes away my sin debt without me having to pay it. Then, as has also been noted, I am justified in God's sight. Because of the righteousness of Christ that is given to me, God sees me as if I was as perfect as Jesus himself. The judge of all the earth (God) says in the court of Heaven, "This person is just. He is righteous in my sight."

Salvation also allows me to be adopted by God as his true son or daughter. Now that there is no more sin to come between us, God the Father legally adopts me, and I become a part of his family. This means that Jesus, his Son, is now my brother, and all other Christians, because they too have been adopted into that same family, become my spiritual brothers and sisters.

Furthermore, because I am a part of God's family, I become heir to God's riches. Everything that is his is mine. His kingdom is mine; the earth is mine; all things are mine.

And, more than that, eternal life is mine. When I die, I will live forever with God in a place that knows no sin, suffering, or pain.

All of these wonderful things happen when I am saved.

Baptism

After a person completes the steps of salvation, one of the first things he should do is be baptized. You have probably seen a baptism on TV, or maybe you have attended one at a church. When you are baptized, another disciple dips your entire body into the water and pulls it out again. This is usually done with other believers looking on. The disciple performing your baptism (it can be a preacher, but it doesn't have to be) usually says something like this before he lowers you into the water, "I baptize you in the name of the Father, and of the Son, and of the Holy Spirit." This statement comes from Jesus' command about baptism in Matthew 28:19.

Before we talk about what baptism does for the believer, let's talk about what it doesn't do. Baptism doesn't save you. Baptism is a symbol of the salvation you have in Christ because of repentance, faith, and commitment (the things we talked about earlier). You do not have to do it to be saved, and you will not be condemned on Judgment Day if you have not. However, it is something every believer should take part in.

Baptism serves at least two important purposes. First, it shows the world you believe that Christ was crucified, buried, and resurrected from the dead; the act of going into and coming out of the water symbolizes this. It also shows that you believe you are now dead to yourself and your old way of life, and that your life moving forward will be lived to serve Christ; when your body slips below the surface of the water, you, in a sense, are entering the grave (a symbol of death), and when you come out, your life, in a sense, begins all over again with a new purpose. Because of what baptism symbolizes, it is not something you should repeat over and over. You should only do it once.

CPSIA information can be obtained
at www.ICGtesting.com
Printed in the USA
FSHW022032200421
80677FS

9 781637 959862